THE BLACK DIARY

More from Nick Redfern and Lisa Hagan Books

Men in Black: Personal Stories and Eerie Adventures

Women in Black: Creepy Companions of the Mysterious MIB

The Roswell UFO Conspiracy:
Exposing A Shocking and Sinister Secret

365 Days of UFOs: A Year of Alien Encounters

Published by Lisa Hagan Books 2018

www.lisahaganbooks.com

Powered by

SHADOW TEAMS

ISBN: 978-1-945962-11-0

Cover design and interior layout by Simon Hartshorne

THE BLACK DIARY

M.I.B., Women in Black, Black-Eyed Children, and Dangerous Books

NICK REDFERN

CONTENTS

"Some things are best forgotten if they can be,
and certainly not set down in a book"

Dr. Philip Gosee

author and naturalist, 1935

INTRODUCTION

Even for a seasoned author like myself – someone who has written more than forty books across the last two decades – the publishing world can still present challenges. Deadlines, edits, fact-checking, securing photographs to accompany the text, and planning publicity campaigns, can all be time-consuming at times. In the end, however, it's all worth it.

That said, there is another aspect to book-writing which can present the author with more than a few challenges. And, on more than a few occasions, those same challenges are steeped in menace and supernatural activity. At least for me they are. Perhaps for you, too.

The late and acclaimed John Keel – the author of two of the most important books on paranormal phenomena, *The Mothman Prophecies* and *UFOs: Operation Trojan Horse* – quickly came to realize something ominous. When you immerse yourself in – and write about - the realms of the unknown, the "things" that inhabit those same realms are quickly driven to intrude upon *your* personal space. And, given the chance, they'll screw you over, left, right, and center. Or, as "Alexandra Leek" - a Keel-based character in the 2002 movie version of *The Mothman Prophecies* played by the late Alan Bates - said: "When you noticed these things, they noticed that you noticed them."

I could not have put it better myself.

I have experienced this strange and eerie situation on far more than a few occasions. For the most part, I find it puzzling and intriguing –in equal measures. Occasionally, however – and despite the fact that I have a gung-ho approach to life and I am not easily intimidated or disturbed by what lies on "the other side" – I have found myself in situations that range from the diabolically deranged to the downright insane. And, there is no better example of that than an approximately three-year-long period – from the summer of 2014 to the final months of 2017 – when I was immersed in promoting my books *Men in Black: Personal Stories and Eerie Encounters* and *Women in Black: The Creepy Companions of the Mysterious M.I.B.* And writing this book, too, of course – the third and final title in a trilogy of sinister tales of the MIB and their cohorts. It was as if by writing, speaking, and even *thinking* about the MIB and the WIB, I was practically inviting them into my reality – which is, in essence, something that is at the heart of this book. And it's also exactly what happened.

We're talking about occult backlash, synchronicities of the jaw-dropping kind, weird phone-calls, and ominous runs of bad luck that appear to have been orchestrated by things that are foul and malignant. Not only that, there were shadowy figures fleetingly manifesting in my apartment in the dead of night, and what I fully believe to have been the "invasion" of my dreams by dangerous entities. Even the "infection" of others who were on the peripheries of the events - such as my literary-agent, Lisa Hagan, whose company has taken what might well be a definitive risk by publishing the very book which you are now reading. And, on more than a few occasions – and despite the undeniable cliché – it really *did* all go down on the proverbial dark and stormy night. Or, to be completely accurate, on dark and stormy *nights.* I seem to be hit by a lot of those. A hell of a lot of them.

I have chosen to share my stories and experiences with you – not to mention all of the mayhem and mystery that went down – for several reasons. Firstly, you may have experienced something very similar to me; in which case, I hope my words may offer you some degree of relief. On the other hand, those same words may only amplify your anxieties; time will possibly tell.

Secondly, knowing that you're not alone in having to deal with such infernal phenomena may provide you with some degree of peace of mind. In other words, no, you are not going crazy. Unfortunately, it's worse than insanity could ever be: *you may well be under unrelenting supernatural attack.*

And, thirdly, I'm here to warn you – and to warn you to the absolute best of my ability – that *opening* doors of the occult variety is a relatively easy thing to achieve, whether deliberately, consciously, or even accidentally. *Closing* the doors to the non-human things that so relish crossing the veil when called forth, however - or even as the mood takes them - is no easy task. In fact, it's nigh on impossible. Unless, that is, you're aware of certain procedures designed to forever banish their icy presence. But, we'll come to all of that later.

The Black Diary is a cautionary tale of what happened when I went looking for – and wrote about - the truth of the Men in Black, the Women in Black, the Shadow People, and the Black-Eyed Children, who, in my view, are all part and parcel of the same thing, whatever that thing may be. I found not just answers, but also unearthly phenomena and terrible creatures that seem to thrive on nothing but grim and wretched negativity and turmoil and chaos.

In light of the above, if I were to tell you to burn this book after you have finished with it, I would only be half-joking. In some of my darker moments, there's absolutely no joking about it, at all. That you're now holding it in your hands, or reading it on your Kindle, may mean that it's already too late: you're infected. Just like me. Yes, some books really are dangerous. And when each and every page has been digested, *they* may be coming for you, too…invading…slowly banging or clawing on your front-door in the dead of night…demanding to be let in… *Or else…*

1

"ALL WAS NOT WELL IN ARLINGTON, TEXAS"

July 12, 2014

Anarchy at the Apartment

The weekend began in fine fashion, as it almost always does. On Friday evening – a couple of hours after finishing one of the chapters of my *Men in Black* book, which was published in the following year - I headed off with a bunch of mates to a local pub to watch an English soccer game on the owner's big-screen television. A great deal of beer was drunk, and plentiful plates of fish and chips were devoured. It was just like being in my old stomping grounds in England. With the game over, we headed off to *The Lodge* (an extremely welcoming Dallas-based establishment at which gyrating women remove items of clothing to the musical accompaniment of the likes of Motley Crue and Rammstein in return for $20 bills). After that, it was on to a *Denny's* to soak up the booze with something greasy and carb-loaded. I slept in on the Saturday morning, finally dragging myself out of bed around noon, and then got on with more than a few chores I needed to take care of.

I was all done by around 5:00 p.m. – which is right around the time that my week of madness and chaos began. I sat down to eat a pizza and thought to myself: why is it suddenly so hot in here? I got up off the couch and put my hand up to one of the air vents. Damn! It was coming through warm. And, worse still, the temperature outside was in the mid-90s. I quickly called the emergency maintenance number.

Within about fifteen minutes, Manuel, the chief maintenance guy at my apartment complex, knocked on my front door. After around another fifteen minutes or thereabouts, Manuel broke the bad news: my air-conditioner unit was fried. Fortunately, the repair didn't cost me anything, as it was all covered by the complex's insurance policy. But, Manuel was unable to get everything up and running until Monday. So, he installed a single window-unit in my bedroom, where I was forced to hang out for two days. "Cooped" would be a better term. Each and every attempt to leave the bedroom had me walking into a torturous wall of overwhelming heat. Not fun. Come Monday morning, however, all was good and I had a new unit. *For a few hours, anyway, all was good.* I didn't know it at the time, but things were about to get seriously sinister.

July 14, 2014

Electrified!

Monday evening was in stark contrast to Saturday night: the temperature was still through the roof, but the sunny skies were gone. In their place was a thick, all-dominating blanket of dark clouds. They raced across the skies, propelled by a fierce, howling wind of near-tornado-style proportions. The rain was pouring down in buckets. Lightning flashed violently and thunder rumbled, both to degrees I can honestly say I have hardly ever experienced, before or since. Around 6:30 p.m., I put on a hoody, zipped it up to the neck, and pulled it down tight over my head. With the rain coming down like a Biblical deluge, I raced down the steps of my apartment and ran to the mailbox, dodging quickly growing puddles. On retrieving the mail, I charged back home. I took the fourteen steps two at a time, flung open the door, and slammed it behind me. I was soaked. And then *it* happened.

Barely two or three seconds after I shut the front door, there was an almighty bang; a brilliant white-yellow flash filled the corner of the living room - and right next to the door itself. Had I not taken those steps two at a time, and was I not around twelve-feet into the room, I would likely have been as fried as the fish and chips I heartily devoured a few nights earlier. The door was singed, the room was overwhelmed by the unpleasant odor of burnt metal, and all of my AT&T equipment and my landline were destroyed, as was my DVD player. Thankfully, and somewhat inexplicably, my television escaped utterly unscathed. Given the timing, it was hard for me not to think that the lightning was meant for me. Based on what happened next, it just might have been.

The next morning was spent buying a new landline and a DVD player and then hanging around, awaiting the AT&T guy's arrival. No sooner had I installed them when *something else* happened – my cell-phone stopped working, and for around five hours. I took the battery out, replaced it, turned it on and off – all to no avail. I was about ready to smash it against my office wall when, quite out of the blue, it started working again. It has not given me any hassle since. My trials and tribulations were still not over, however.

July 16, 2014

Attacked from Above

Midway through the following afternoon I took a casual stroll down to the mailbox, which was in stark contrast to my actions on the night of the storm. As I crossed the car-park area, I caught the sight of something dark and compact hurtling towards me: It was a blackbird. The creature, for reasons I still cannot fathom, launched a full-scale attack on me, and just missed my face as I ducked down. I spun around and saw the bird propel itself to a height of around fifty feet. I stared for a moment, noting how it seemed to hang in the air for a second or two, and watched with a mixture of trepidation and "what the fuck?" as the bird launched another attack.

It dive-bombed me, screeching and squawking in deranged fashion, and slammed into my back. It clung on by its claws, as its beak repeatedly pecked my shoulder blades in mad fashion. Only when I managed to twist my hands behind my back and tried to grab it did the beast cease its attack. As I tried to grab it, it made the craziest noises I have ever heard. It then flapped its wings and vanished at high speed. To say that it was all very reminiscent of Alfred Hitchcock's classic movie, *The Birds*, would not be an understatement. In fact, it would be right on target.

Barely avoiding getting hit by lightning, the destruction of an air-con unit and a landline, a DVD player, my AT&T equipment, and the temporary breakdown of a cell-phone, and all within a few days of each other - coupled with not just one violent bird attack, but *two* – led me to suspect that all was not well in Arlington, Texas. Something evil was afoot; something dangerous and deadly. That became even more graphically clear two days later.

July 18, 2014

Buffy, the Djinn, and Another Lightning Strike

Shortly after 10:00 p.m., I got a phone call from a good friend of mine, Buffy Clary. She's a gifted psychic who lives a few hours drive away, in Tyler, Texas. She had something startling and disturbing to reveal. In the very same week, she, too... *had been struck by lightning.* Making things even more ominous, Buffy had just immersed herself in the dangerous domain of the Djinn – after I loaned her my copy of Rosemary Ellen Guiley's book on the subject, *The Vengeful Djinn*, which, with hindsight, was probably a most unwise thing for me to do.

By that, I don't mean Rosemary's book is a bad one. In fact, the exact opposite: it's an excellent study of Middle Eastern plasma-based entities that are as dangerous as they are feared. Shape-shifting creatures, they are known to take on multiple forms, including those of the classic Men in Black and fiery-eyed black dogs of the kind which inspired Arthur Conan Doyle to write his almost legendary novel, *The Hound of the Baskervilles.* But, as I noted in the Introduction to this book, just focusing on certain, supernatural phenomena can cause them to manifest in your reality. Almost certainly, that's what happened with Buffy: reading about the Djinn resulted in her getting a visit from them. I even wondered if the act of loaning Buffy the book provoked the week of madness that hit me. The weirdness for Buffy was not over...

July 31, 2014

Texts and the Tall Man

Just before noon, Buffy texted me a story with somewhat MIB overtones attached to it. She wrote: "During a reading yesterday, I pictured a tall man in a grey suit walking out from the light of the street lamp holding his hat down so I couldn't see his face."

About twenty minutes later, she sent a second text: "It had a London-cobbled street-Jack the Ripper feel but I didn't mention that because thought it was too cliché."

As you'll come to see, Buffy is someone who pops up regularly in the pages of this book, always in relation to matters supernatural. Here, the story is more a fragment, but Buffy has encountered more than her share of weirdness, as you'll see.

2

"THE MAN SPORTED A WEIRD GRIN"

September 2, 2014

"He was met by three very pale men dressed in dark suits"

History is made in many and varied ways. For "Harry Palmer," it occurred deep below Wright-Patterson Air Force Base, Dayton, Ohio in the 1980s. I know this, as Palmer contacted me on a dark September 2014 night to share a very strange story – and after reading my 2012 book, *The Real Men in Black.* As Palmer revealed, he had debated for weeks as to whether or not he should take the plunge and speak to me. On the one hand, he wanted to get my opinions on what it was that he encountered. But, on the other hand, he was somewhat concerned about the possibility of his revelations causing "troubles" – specifically with Uncle Sam.

It was around 11:00 p.m. when we started chatting and it was well into the early hours when we finally wrapped things up. I spent the entire time on my back, stretched out on the couch and listening carefully. Although Palmer was somewhat hesitant at first, he soon gained his wits and opened up. What follows is a summary of the strange and sinister experience found himself in, more than a quarter of a century earlier.

Wright-Patterson AFB is a highly secure military installation that has a reputation for allegedly being the home of a number of well-preserved – and some *not* so well-preserved – corpses of dead aliens, presumably recovered from more than a few UFO crashes. Many UFO skeptics ignore or write-off such claims. Harry Palmer knows better.

It was a winter's night in 1988 when Palmer was ordered to report to one particular building on the base that he had never previously been in, one which was connected to a certain weapons-storage area. On doing so, he was met by three very pale men dressed in dark suits; they directed him to a door which, when opened, revealed a large elevator on the other side. Silently, the black-clad trio motioned him into the elevator. He quickly realized he was descending – and to a fairly deep degree. He was then ushered into a corridor, which had a large vault-like door at its end.

One of the three men opened it, and in a strange high-pitched voice, ordered him into the vault. The same man pointed at a large container – perhaps nine feet in length and five feet in width – and ordered Palmer to take a look inside. He did as he was told and was shocked – to the point of feeling nauseous and clammy – by the sight of a badly damaged body of what he, Palmer, could only guess was an extraterrestrial: The head was large, the eyes were huge and black, and the severed torso was skinny.

In seconds, Palmer was forcibly taken from the room, then taken to yet another room, and then ordered to sign a document that effectively said that if he ever spoke of what he saw he would be prosecuted to the extent of the law for violating U.S. national security. After signing the document, a deeply worried Palmer was taken back up the elevator and unceremoniously left there to make his way back to his normal place of work.

Clearly, this affair makes no sense at all – unless someone was playing weird mind-games with Palmer, although for what bizarre reason one can scarcely guess. After all, why even show him the body in the first place? His work at the base had *zero* to do with UFOs. The whole thing, I told Palmer, seemed like some bizarre theatrical event, but for whatever purpose was anyone's guess. Palmer agreed.

There was a curious afterword to this particular story: The very day after Palmer contacted me, he received four hang-up phone calls in the early hours of the morning. It was something that led him to regret sharing his experience with me, as he admitted, when he called me back the following night. That I told him hang-up calls in the dead of night – and specifically in relation to UFO issues - were typically attributed to the Men in Black, didn't really help him to relax. How could it?

"Don't use my real name if you publish this," a worried Palmer said, before hanging up with not even a solitary "goodbye." Yep, another weird night in almost a lifetime of them.

After speaking with Palmer for the second and final time, I had to wonder what, exactly, a group of clearly non-human MIB were doing in the depths of Wright-Patterson Air Force Base on that mystery-filled night in 1988? Were they working alongside the military? Have the MIB infiltrated the U.S. Government? If so, which side is in control? Is *anyone* in control? It was questions like these that made me wonder just how weird the MIB issue really was – and still is.

September 9, 2014

MIB dreams

A week later, yet another MIB report came my way. Like most such reports involving the black-suited things that have caught my attention, it was a bizarre one, and one which had significantly traumatized the witness. Only hours after briefly seeing a cigar-shaped UFO hovering near Stonehenge, Wiltshire, England on December 25, 1985, Sandra Green – who was driving home, in the early hours, after a late night Christmas Eve party in the English city of Bath – had a bizarre dream. In it, three Men in Black warned Sandra to keep her nose out of UFOs. They were – as they usually are – cadaverous "slightly dead"-looking MIB. As for the warning, it was one of those "or else"-type threats that the MIB are so skilled at making. The dream was so graphic it led Sandra to believe that the MIB had the ability to literally get inside her dreams, manipulate those very same dreams, and issue a bone-chilling warning – one which she has still not forgotten.

Sandra shared the story with me on the evening of September 9, 2014 – the facts of the affair still firmly fixed in her mind and her memory after close to thirty years. And, I have to say, it wasn't just the facts that were fixed in Sandra's mind: It was cold fear, too. Even across a transatlantic telephone call I could sense that all was not well.

I messaged Sandra a few days later in an effort to clarify a couple of things she mentioned to me. She did not reply. She *still* hasn't. A sign of something? I hoped not.

October 8, 2014

Beware of the Hat Man

The Hat Man is a Man in Black-type being that appears in shadowy form; not unlike the infamous Shadow People – to whom the Hat Man is almost certainly related, even if we're not sure why. On many occasions, however, the Hat Man appears in regular, human form, with a black beard and a black suit, and wearing a long overcoat or a cloak –always black, too. I know this, as I have investigated many such affairs. Most noticeable about this creepy figure is, of course, his hat. Sometimes, it's a fedora, other times it's an old style top hat. Occasionally, it's more like a cowboy hat. But, regardless of the kind of headgear, it's always present and always black. Many of the encounters occur while the victim is in a distinct altered state – that of sleeping.

Angie had just such an encounter in Leominster, Massachusetts, on September 6, 1994, as she told me late on the evening of October 8, 2014. Her reason for contacting me was due to the fact that, quite out of the blue, she had suddenly begun to dream of the events of twenty years earlier – and dream of them almost every night for the last couple of weeks. Very much like those people unfortunate enough to be confronted by the somewhat similar Slenderman – the ultimate Internet meme come to life in Tulpa form. (A Tulpa is a thought-form, a Buddhist term which refers to a creature created in the depths of the human mind and which has the ability to leave the mind and live independently of its creator. In other words, what we imagine, can supernaturally come to life.) Angie was sleeping when the mysterious thing disturbed her sleep by manifesting in her bedroom and staring at her with a menacing grin on its pale, ghoulish face. For a few moments, Angie was unable to move. As she finally broke the spell of paralysis, however, the Hat Man was gone.

I was hardly surprised when Angie added that, as a result of the new dreams, she now feared a return visit from the man in the hat. I tried to stay on the positive side of things and told Angie to try and dismiss such worries from her mind. But, to this day it would not surprise me if Angie gets that second feared encounter, as my words really did not seem to comfort her, even when we chatted again, some weeks later on.

November 9, 2014

Don't say a word

The Marshall family of Perth, Australia had a strange encounter after seeing a huge, black, triangular shaped UFO flying completely silent, and at a dangerously low level, as they headed home after a night out on June 3, 1999. Although they did not tell anyone about their experience (at least, not at the time, anyway), late on the following night the phone rang. It was a man with a strange and somewhat European accent who warned them not to talk about the craft they had seen. In this case, the family forgot about the experience until years later when they saw a TV documentary on the Men in Black enigma - which referenced the issue of the MIB making threatening phone calls to UFO witnesses. Today, the family is solidly convinced that what they experienced was full-blown intimidation from a menacing MIB. I had to agree, when I read the facts, which came to me through Facebook on November 9. It was yet another chilling affair to add to my files.

November 22, 2014

"The government is concerned"

Gloria is an elderly woman who lives in Decatur, Texas, who I met with on the afternoon of the 22nd. I decided to make the approximately 110-mile-roundtrip after hearing a bit of her story down the phone the previous evening. I set off early, wondering as I always do, what exactly I might be in for. It was a typical November day in Texas, cool and bright, when I hit the road. It could have been a less than extraordinary experience – as is sometimes the case. But, not this time. On this occasion, the trip was well worth it. I arrived at an old house – probably dating from the 1940s, and which was well kept and had a welcoming porch with a couple of chairs. I knocked on the door and in just a few seconds it opened. In front of me was Gloria, a white-haired lady who smiled broadly. At least, her experience wasn't affecting her character, I thought. She invited me in and I sat down, as a couple of caged canaries bid me welcome. At least, I think that's what they did.

On July 19, 2012, Gloria told me – as we drank coffee and ate homemade lemon cake in her living-room – she briefly saw what can only be accurately described as a flying saucer, which hovered over her home as she sat in her backyard, reading a book and with her two dogs for company. In fact, it was the barking of both dogs - which stared intently and rigidly at the sky - that alerted Gloria to the presence of the weird craft. It didn't stay around for long, however. It was a case of here-one-second-and-gone-the-next-second. But that was not all.

The next afternoon, that of the 20th, there was a knock at the door. It was a pale-faced, thin woman of about thirty, wearing a long black wig and dressed in a black jacket, a white blouse, and a flowing, black skirt. And then there were the huge sunglasses. And the WIB smelled of dirt – something I have heard before. Gloria felt deeply uncomfortable as, upon opening the door, the Woman in Black proceeded to warn her not to talk about the UFO she had encountered the previous day, due to the claim that "the government is concerned." Concerned about what was never explained.

Clearly, the WIB was *not* from the government. Or, from *any* government. According to Gloria, the woman didn't even look human. "Skeletal" would have been a far better description. After asking what the time was, the WIB turned, walked down Gloria's driveway and vanished. Never to be seen again. It was a familiar scenario – one that I knew only too well. I still do.

Gloria thanked me for offering some thoughts and advice on the affair – such as try and put it all behind you, as these things thrive on our fears – and gave me a plentiful supply of that delicious lemon cake to take back with me. We still keep in touch. I'm pleased to say that her WIB has not returned. So far...

January 3, 2015

The man who grins

In 1987, the Maxwell family spent a week vacationing in and around San Francisco, staying with friends in Menlo Park. On their way back home, they traveled along California's famous Highway 101, which provides a panoramic view of the Pacific Ocean, and for mile upon mile. They chose to drive through the night, when the highway would be at its least busy, thinking that it would be to their benefit to do so. How completely and utterly wrong they were. As fate would destine to have it, after a couple of hours of driving, the family of four spotted a strange light in the sky. It was described as a bright green orb of light, about the size of a beach-ball, one which paced their car and that stayed with them for a couple of miles, at a height of around sixty feet. There was nothing frightening about the encounter. Rather, they were all amazed and excited. It wasn't long, however, before things got very disturbing.

The day after the Maxwell family got home was a Sunday, meaning they had an extra day before returning to work and school. It was while one of the teenaged children sat on the porch playing music on an old Walkman that she caught sight of a man on the other side of the road. He was dressed completely in black, aside from a white shirt. He even wore black gloves, on what was a bright summery day. The girl was particularly disturbed by the fact that the man sported a weird grin and was staring right at her. So unsettled was she that she went back into the home and told her father of what had just happened. He quickly went to the door but – no surprise – the MIB was gone.

When Mr. Maxwell told the story to me – on the phone, on January 3 – the anxiety in his voice was as clear as it was eerie to hear. I have seen and heard that kind of anxiety so many times. *Too* many times, I would suggest. And, regardless of the particular location, almost always in relation to the Men in Black.

March 28, 2015

A disturbing incident is revealed

Pasadena, California was the site of a strange encounter with a definitive Man in Black, specifically on March 22 1979. The witness, Charlie H., contacted me thirty-six years later to share his story, after I spoke on the MIB enigma on a local, Texas radio show. Charlie, now living in the Lone Star State, had seen a UFO as he drove near to what is known locally as Devil's Gate Dam. It's a place with a great deal of paranormal activity attached to it and at which Jack Parsons, a 1930s rocket-pioneer and a devotee of "the great beast" Aleister Crowley, hung out on a regular basis. The UFO, said Charlie, was not particularly large, and was circular and bright pink in color. Charlie, who was driving home from a shift that ended at 2:00 a.m., added that the UFO came close to his car – around eighty or ninety feet away – then shot away into the sky.

Two days later, and as he happened to look out of his living-room window, Charlie saw a man dressed in a black fedora, black suit, black trench-coat, white shirt, and black tie, get out of an old, black Cadillac and quickly take a photo of his home. The MIB then got back into the vehicle and drove away. John Keel termed this particular brand of MIB as "phantom photographers." A most apt term, to be sure.

I shared my thoughts and opinions as we chatted and thanked Charlie for his call. We'll hear more about those creepy characters with cameras later on.

April 8, 2015

The boy in black

Time and again I hear stories of the Men in Black and the Women in Black. On February 1, 2015, however, Marc, a hiker, encountered in California woods a *child* in black. Not a Black-Eyed Child of the type diligently pursued by BEC expert David Weatherly, I should stress, but a pale-faced boy attired in black pants, black t-shirt, and a black hoody. The CIB turned up while Marc was hunting, and just a few hours after he encountered a globe-like UFO over his tent at a height of around eighty or ninety feet. According to Marc, who reached me by Facebook, the boy in black appeared out of nowhere, laughing in an odd and hair-raising fashion, and then suddenly vanished into nothingness. I was not at all surprised when Marc told me he didn't stay in the woods for much longer: He was soon gone.

May 23, 2015

The bedroom invader

Point Pleasant, West Virginia is – beyond any shadow of doubt – noted most of all for its wave of sightings of the infamous Mothman, between 1966 and 1967. In 2014, however, the town played host to something even stranger. The witness to the weird affair was a local woman named Denise, who emailed me the details on May 23. As Denise explained, she was jolted from her sleep by the sight of a young boy looming over her bed. This was no normal boy, however: it was one of the dreaded Black-Eyed Children – pale-skinned, black-hoody-wearing kids who are noted for their completely black eyes.

Denise tried to scream out but her vocal-chords were paralyzed, as was her entire body. The eerie boy stared at terrified Denise for a few moments, then retreated into the shadows of the room and vanished. It was a nerve-jangling experience that Denise has not forgotten. Nor has she forgotten a strange wave of hang-up phone calls that occurred across the next three nights, and all around 3:00 a.m. A connection? Denise believed so. As did I.

Two days later, things got *really* weird...

3

"SHE REFERS TO THESE FORCES AS 'SOURCE'"

May 25, 2015

"Challenger exploded"

The day began for me as it always does from Monday to Friday. I got out of bed around 7:00 a.m., showered, had breakfast, checked my emails and Facebook messages, and then got to work, quitting at 5:00 p.m. The day was destined to change from the norm, however, and drastically so. On the last weekend of May – in fact, just a few days later - I was due to speak at the annual *Contact in the Desert* gig at Joshua Tree, California. The subject of my lecture: the UFO Contactee phenomenon. One of the things I planned to speak on (and which I *did* speak on) was the matter of a certain Contactee; a woman who had an intriguing, but very controversial, story.

For those who may not know, the Contactees first surfaced in the early 1950s, claiming contact with long-haired space aliens who demanded we get rid of our nukes and live in peace and harmony. In today's world of Ufology, the Contactees have been largely elbowed out of the window by the Abductees. But, not entirely – which brings me back to the matter of the woman referred to above.

Back in 1986, she was interviewed by the FBI, as part of the Bureau's investigation into the January 28, 1986 explosion of

NASA's *Challenger* Space Shuttle. It was a tragic event that killed all of the crew and created shockwaves around the planet. Declassified under the terms of the Freedom of Information Act, the FBI's files on the incident record that the woman (whose name is deleted from the available documentation – although it is known to me) said that the Shuttle was sabotaged. According to the heavily-redacted documentation which the FBI has been willing to release into the public domain, a source of some standing, one who was apparently very, well known to Bureau agents, had come into contact with the woman. And matters only escalated from there.

The FBI's papers state that the woman:

> ...claims to be in contact with certain psychic forces that provide her with higher information on selected subjects. She refers to these forces as "Source" and when providing information from Source she often speaks in the collective "we." [The woman] claimed that she had come to Washington, D.C. to provide information concerning the Challenger Space Shuttle explosion on 1/28/86.
>
> On 2/24/86, [the woman] was debriefed concerning the information she wished to reveal. The enclosed audio tape [Note: which the FBI has not yet declassified] is a record of that entire session and is self-explanatory. The following points were emphasized during the course of the briefing:
>
> (A) The Space Shuttle explosion was not the result of a technical malfunction. Rather, it was an act of sabotage perpetrated by a terrorist organization.

(B) There were three saboteurs on the scene at Cape Canaveral. These included two ground crewmen and one of the astronauts who died in the explosion.

(C) The terrorist organization is a Japan-based fanatical group with an ancestral lineage and a deep seated hatred toward the United States. Its motivation in destroying the Shuttle was to embarrass and discredit the United States and to impede the progress of the space program.

(D) The three saboteurs were all of Asian or Far Eastern heritage. They were apparently recruited by the terrorist organization after their employment by NASA.

(E) The explosion was effected by a device placed inside the external fuel tank of the Shuttle. An individual whose description seems to match that of an engineer or technician placed this charge. The charge was triggered by a second saboteur using a hand-held transmitter while standing in the crowds watching the Shuttle liftoff. The individual matches the description of a guard or security person.

(F) The astronaut saboteur chose to die in the explosion as a sort of ritual death or "cleansing".

(G) The descriptions of the saboteurs provided by [source] are probably complete enough to pinpoint the individuals. Although names and addresses are not available, in the case of the 'guard' [source] pinpointed his residence on a 1:24,000 scale topographical map of an area of Florida near Cape Canaveral. She also provided a description

> of his physical appearance as well as his life style. While actual names are difficult for source to acquire [source] claims she could easily identify the saboteurs from a list of names if NASA could provide one.
>
> (H) Source's motivation in revealing this information is to assure the United States that its space program is a safe undertaking and that there is no major technical flaw in the Shuttle equipment...Source is concerned about possible lengthy delays in the progress of the space program that will allow other nations to surge ahead. [Source] predicted that wreckage of the Shuttle will be found within the next several days that will support the sabotage allegation.

In essence, that is the extent of the relevant, declassified material on this particularly unusual, highly controversial, and probably unique conspiracy theory. Even though some may scoff at its contents, it's notable that the FBI has admitted to withholding in their entirety, and in the specific name of national security, no less than 26-pages of secret documentation on this weird saga of the allegedly sabotaged Space Shuttle.

Well, given that the woman claimed contact with supernatural beings, I decided to reference all of this in my Contactee-themed lecture at Joshua Tree. So, I surfed the Net for a good picture of the ill-fated *Challenger* that I could insert into my PowerPoint presentation. A couple of hours later, I had to make a call to my agent, Lisa Hagan – on the matter of none other than my then-forthcoming book, *Men in Black: Personal Stories and Eerie Adventures.*

During the course of the conversation, Lisa said that something very strange had happened to her earlier that same day. I asked what it was and she told me how she received a phone call – number not available – from someone who spoke just two words: "*Challenger* exploded." Lisa is one-hour ahead of me, and when we checked, it turned out that she got the phone-call right around the time that I was searching online for the photo. It scarcely needs to be said that Lisa was amazed - and quite a bit unsettled, too - when I told her I was looking into the *Challenger* issue the very morning she got the call. I had a deep suspicion – one that has not gone away – that my every move was being monitored; probably Lisa's, too. After all, both of us knew all too well that the MIB and mysterious phone calls go together like hands in gloves.

July 16, 2015

Shattered!

For me, it's always a good day when I put the finishing touches to a manuscript. And, June 16, 2015 was the date on which I completed the writing of my book, *Men in Black: Personal Stories and Eerie Encounters.* Of course, the manuscript still had to go through the editing and proof-reading stages. And I had to dig out a bunch of photos to go with the book, too. But, there's a good feeling of satisfaction, while sitting at one's desk, knowing that the bulk of the work – in terms of the content of the book – is done. Plus, it was a hot and sunny day, the birds were singing in the trees outside my apartment, and that night I was due to see a Motley Crue tribute band play in a local beer and junk-food dive. It was all good! Except, that is, for one thing which occurred around 9:45 a.m. That was roughly the time when I made the final change to the manuscript, hit "save," and closed the document. I was ready to email it to my agent Lisa for review.

Only seconds after I closed the document, I heard a sudden bang coming from one of the rooms in my apartment. I frowned, stood up, and probably said something to myself along the lines of "WTF?" Since my apartment home is a relatively compact one, it didn't take long at all for me to find the cause of that bang; seconds, in fact. On walking into my bedroom, I saw that one of the many pictures I have on my walls had fallen to the floor. Despite the floor being carpeted, the black picture frame was broken and the glass had shattered, with pieces and shards all over the carpet. Dammit. Time to get out the vacuum. What was particularly eye-opening, however, was the specific item which had fallen from the wall.

It was a framed letter written back in 1953 by none other than Albert Bender. He was the man who, arguably, birthed the mystery of the Men in Black. It was all as a result of his early-1950s-era traumatic experiences with a trio of glowing-eyed, vampire-like MIB in his hometown of Bridgeport, Connecticut. The strange tale was chronicled in Bender's 1962 book, *Flying Saucers and the Three Men.* It wasn't long after that initial encounter in the fifties that Bender, who had got into UFOs in the late-1940s, and who created the International Flying Saucer Bureau, couldn't take anymore mayhem and menace in his life and quit Ufology for good. He did not look back. Well, maybe a glance or two, but certainly not much more.

At the time the frame hit the floor, the maintenance guys at the apartments were working outside, hammering away at something. So, one *could* make a case that the vibrations from their tools dislodged the picture and – hey, presto - I'm left with a broken frame and glass everywhere. Maybe that's all it was. On the other hand, though, it's worth noting that the wall in question has around fifteen or twenty small-framed images hanging from it. Of all the ones that could have fallen (and which range from images of Bigfoot to the Loch Ness Monster, and from the Chupacabra to 1950s über-babe Betty Paige, and much more), it was one with a direct connection to the Men in Black. I could hardly fail to note that it fell at the same time I not only closed the Word document, but when I had just completed the *Men in Black* book, too. I got a distinct and sudden feeling that invisible manipulative entities were watching – and coldly basking in – my every move.

When I told a few people about this, they all said that, in essence, it was a sign. But, a sign of what exactly, was the important thing that no-one could fully agree upon. There were those who viewed it as a warning to me to stay away from the matter of the MIB. Others suggested I was under some kind of dark demonic attack. My view? It was just another day of high-strangeness; I shrugged and made a plate of toast and jelly. There was, however, one thing I knew for sure: Based on what had occurred over the course of the last year, it was likely to be just the start of yet another bout of paranormal madness and mystery. Sure enough, it was that. And more. As always, though, I don't ever let such issues affect my life. The Motley Crue tribute gig went ahead without a hitch. The band was pretty lousy though.

August 16, 2015

MIB invade the airwaves

Although this book chiefly chronicles events that occurred between 2014 and 2017, it's worth noting the following, before I get to the story I am about to tell – as there is a high degree of relevance. Back in 2011, when I was promoting my second book on the MIB, titled *The Real Men in Black*, one of the things I was keen to discuss in radio interviews was the matter of MIB and telephone interference. We're talking about strange voices on the line, weird electronic noises, and hang-up calls. Several people contacted me to report they were experiencing the exact same thing – but only *after* they had read my book. *Before* reading it? There were no problems. Not a single hitch.

This was very much like certain issues that surfaced with regard to David Weatherly and his *Black-Eyed Children* book, and from the time of its publication in 2012. It was as if when people read my book (and David's too), it triggered paranormal activity in the home of the reader. Yes, I know how wild that sounds, but it truly is what happened. Of course, I couldn't fail to note that this also mirrored what happened to Buffy Clary when, on July 18, 2014, she dug deep into Rosemary Ellen Guiley's book *The Vengeful Djinn*: She was electrocuted. Thankfully, not seriously – but certainly enough to scare her half to death.

Indeed, telephone problems were absolutely rife throughout that period. As an example, Whitley Strieber invited me on his *Unknown Country* show to talk about my MIB book and we experienced endless amounts of very strange interference and odd sounds on the line. The exact same thing happened, at a later date, with *Coast to Coast AM* – something which took on even stranger proportions when my landline phone suddenly quit. So, I proceeded to use my cell-phone as a back up. The battery drained astonishingly quickly, despite being fully charged. And, so, I had to hastily dig out of a cupboard an old, spare, cordless landline phone, which finally allowed us to complete the show. See what I mean now about dangerous books? They are everywhere.

There was something else too: These odd situations only revolved around my books on the MIB. For all of the other interviews, on other topics, not a problem was in sight. But, with the MIB, *endless* problems. And no, this was *not* some odd publicity campaign; certainly, I can think of far better, and much easier, ways to promote a book! In light of all this, I should have known that when I decided – for the *third* time – to write a book on the Men in Black, and then one on the Women in Black, too, I would find myself plunged into a world of menace. All of which brings me to what happened on the night of August 16, 2015.

I was on Steve Warner's *Dark City* show, which operates out of New York. Of course, we were discussing the MIB. During the interview, Steve experienced something very odd. It revolved around the lights in his home. Namely, they were repeatedly turning on and off. I don't mean flickering on and off, as they might during a violent storm. Rather, Steve - sat in front of his computer – could hear the light-switches being *turned on and then off* – as if by invisible "hands." Maybe claws or talons would be far better candidates. It was something that actually became an entertainingly weird part of the interview!

When I mentioned all of this in a two-part online article at *Mysterious Universe*, it generated a few replies, including one from "Rob from Rigel VII," who wrote: "You may not believe this, but as soon as I pulled up part 2 of this article, the power went out at work, which is a large office building. The UPS boxes started beeping and then the power all came back. Eerie!"

Yep, very eerie. But, somehow, not at all unexpected.

September 14, 2015

A reader speaks

One of the people whose MIB experiences feature in my 2015 *Men in Black* book is a friend of mine, Steve Ray. After reading the book, and just a few days after it was published, Steve experienced something unusual. In his very own words to me: "I read the whole Kindle version of the book Sunday and Monday.

Monday night I came home and found two black cars parked – headlights out – in non-assigned spaces directly facing my assigned parking space. When I came upstairs, I found my living-room lamp had been switched from its normal setting to the spookier black-light setting – which I have no memory of doing, and I was the last one here."

I told Steve I was not surprised. He was not comforted by my words; not at all. He wasn't meant to be. I tell it as it is: good, bad, or worse still.

4

"'I CAN HELP YOU. JUST SAY *YES*,' IT SAID"

November 21, 2015

"Now, my hands are shaking as I type this"

John Winterbauer is a friend of mine who I run into about once a year or so at conferences. At 8:43 a.m. on November 21, John sent me a text - while clearly in a state of some concern - that revolved around the story of a woman named Terry. Her MIB-themed encounter is related in my *Men in Black* book of 2015. In part, Terry had told me:

> This encounter occurred during the day time thirty-five years ago in 1977 at a Winchell's in Sunnyvale California. I was just sixteen years of age at the time. My best friend and I had just finished our coffee and were heading for the doorway to leave. As we were passing, a man seated at one of the tables reached out and firmly grabbed my wrist. In that moment, my first reaction was to jerk away from his grip, but looking down at him I realized that he was a very old man and I felt in that moment that he was harmless. He was dressed in black wearing an old- fashioned hat and suit, his clothing looked like it was from the 1930's. He was extremely pale, very thin and appeared to be very old, I guessed in his 80's or 90's.

He told my friend and I that he was a palm reader, a very good one, he claimed and could he please do us a favor by giving us each a reading. My friend and I talked briefly to each other about his offer and agreed to let him. I first sat down across the table from him, then as my friend was starting to sit down along side of me, he stopped her and asked her to please go sit a few tables away out of earshot from us, explaining that the information was going to be direct and personal, we would need some privacy. She complied and moved several tables away.

Once my friend was seated away from us, I placed my hands on the table in front of him, palms up and looked into the man's face, it was then that I noticed the old man's eyes were completely glazed over with cataracts, he was surely blind I thought, as his gaze was unfixed and unfocused in my direction. I asked him how long had he been a psychic? He responded by saying that his ability had nothing to do with being psychic and emphasized that it had everything to do with 'science.'

He began by talking about my early childhood experiences along with some very painful incidents that had occurred to me in which he had precise and detailed information that I was sure no one could have known about. Quite suddenly I felt very vulnerable and exposed as he recounted these events, his knowledge unnerved me to my very core. Staring straight ahead he moved along into my present situation, lecturing and chastising me like a father would a child, for some poor choices I had made during that time. All the while he broke a doughnut down into tiny pieces. My mind raced as I tried to figure out how on earth could he know this stuff? It was then in that moment that I would forever change my ideas of

> 'secrets kept' and how I viewed my own identity in the world. Apparently 'nothing' could be hidden from myself or anyone else for that matter. I felt naked in the truth."

All of which brings us back to John Winterbauer and his text of November 21. It read:

> Morning, Nick. Odd occurrences last night. I dreamed a woman came to warn me of an unspecified forthcoming trouble. *She identified herself only as Terry* [italics mine]. This morning I'm reading your new MIB book and just finished the donut shop account. My girl and I are getting ready to head to St. Louis for a weekend of paranormal adventure. This dream had no meaning this morning... now my hands are shaking as I type this. Most odd...most unsettling. Probably means nothing but the coincidence has me rattled. Thought I'd share in case I vanish or some weird shit this weekend! For the record I've enjoyed the book up to this last story!

Thankfully, nothing untoward happened that weekend. John has not, however, forgotten this undeniably sinister and thought-provoking synchronicity concerning Terry. Neither have I. I saved that text and - as short as it is - I still go back and read it, now and again.

February 8, 2016

Something comes calling

The day was a very strange one. Or, to be absolutely correct, it was a strange night. I went to bed around midnight, and, at roughly 2:00 a.m., I was semi-woken up by the sounds of what began as unintelligible, disembodied mumblings. They appeared to be coming from something lurking in the shadows of the darkened hallway that links my bedroom to my living room. I then heard something speak to me in a deep, gravelly, and hoarse whisper that was not at all unintelligible: "I can help you. Just say 'yes,'" *it* said.

The skeptics would almost certainly say that what I experienced was a bout of so-called sleep-paralysis – a condition that can result in an inability to move, and a sense of intense and impending danger in the bedroom. I don't doubt that's what it was. But, where I differ from the skeptics is that, unlike them, I believe sleep-paralysis has an *external*, rather than an *internal* origin.

We're talking about dream-invaders of just about the vilest kind possible. Even in my partially asleep state I knew there was nothing but dangerous deception and manipulation at work. I said out loud "No," and focused on putting a barrier between it and me. Whatever *it* actually was. I got a distinct and disturbing feeling that had I said "Yes," I would have given the unearthly thing permission to invade my space and wreck god knows what kinds of havoc and mayhem. I also got the feeling that the thing was massively pissed by my negative response.

February 9, 2016

"Why would someone purposefully invade someone else's dream?"

In light of the "dream invasion" of the early hours of February 8, I spent a great deal of the 9th to the 11th researching, and pondering on, the means by which supernatural entities may be able to invade and take control of our dream states. One of the people whose studies I focused upon was Samuel Hatfield, a mystic and "energetic healer." In his own words:

"Why would someone purposefully invade someone else's dream? Well, there are a number of reasons; the chief among them is to influence someone's thinking. Much of what occurs in a dream is left in the subconscious or unconscious mind. The subconscious / unconscious mind influences much of our daily lives. It governs the automatic responses and processes such as memory, motivation, instinct, and even emotional reaction. Purposeful invaders often seek to influence these things to invoke particular responses in others. They do this by entering the lucid dream, or even pulling someone into their own dream and implanting thought forms to condition the subject."

I have to say that, as I read Hatfield's words, I could not deny this sounded eerily like much of the weirdness I experienced. And, for no less than two more nights, similar occurrences took place. On these occasions, however, the voices were always unintelligible. All I could tell for sure was that words were being spoken – in a rapid-fire, almost wildly mad, fashion. But, what those words were, I had no idea. I *still* don't know. Then, suddenly, the dream-invaders were no more. Perhaps my reluctance to do their bidding led them to turn their evil attentions elsewhere and target some other soul or several. There was still more to come, however.

February 12, 2016

"One evening I kept getting the eerie feeling of being watched"

The very next day I received the following from an Englishman named Liam Robinson:

> I bought your book *The Real Men in Black* off of Amazon a while ago and would like to share an event that happened to me in 2011. This was before I had read the book or heard about the more sinister alien type MIB you describe, but had heard of the government G-Men gathering UFO reports from witnesses.
>
> Here's my experience, it might not be anything relevant but I found it quite interesting. It was either late July or early August and I was on holiday in Weymouth. This was days before the London riots of that year, staying with my parents in a rented accommodation at Whitehead Drive. The room I was in had a lovely view of the bay and Chesil Beach; you could see Portland in the distance. As the view was nice and the weather was hot, I would always leave the window open for air and to admire the view. I had a laptop set up on the desk near the window so could look out easily.
>
> One evening, I kept getting the eerie feeling of being watched. I would look out the window and not see anything of note, the sun was beginning to fade out at this point, I would guess it was about 5-7 p.m. sort of time, but you could still see the road and the beach easily.
>
> Every time I went back to using the laptop, I would again get this horrible feeling of unease and would look

back out of the window. Eventually I spotted what appeared to be a man almost comically hidden behind a road sign. He stood so half of his face was visible from the side of the sign but the other half obscured. His legs were visible underneath the sign, however.

The man was I guess about 200 yards from my window but I could make out he was wearing a full dark suit, which was very odd considering it was a warm summer and everyone else nearby was wearing shorts and t-shirts. He also had a black hat on, I guess either a fedora or a bowler-hat, but it was hard to make out the exact shape at the distance. He didn't appear to be staring at me but rather straight down the carriageway, his back was to Portland so he was facing towards Weymouth. I watched him for a good few minutes and he stood completely motionless, staring forward. The weird thing was none of the cars driving past seemed to slow at all as they passed him.

At this point I quickly ran into another room to grab some binoculars and went back into the bedroom and he had vanished! I could only have been gone for 30-60 seconds max and his position would be hard to get away from unless he jumped into the harbor or got into a car. However the flow of traffic meant that it would be very difficult to get a car to stop, pick him up and start again without causing horns to blare from other drivers.

As I said it could be nothing of note at all, but after reading your book and listening to several radio shows where the Men in Black were discussed, it seems very similar to one of those events. Thanks for reading, Liam.

5

"TWO IDENTICAL, TALL, BLOND MEN IN DARK SUITS"

February 26, 2016

"Two identical, tall, blond men in dark suits"

Samuel Hatfield's words – and my experiences – about strange dreams were made all the more thought provoking by something that happened on this particular date. It was around 8:00 a.m., the temperature was cold and I was in my office, sitting in front of my laptop, having just finished replying to a handful of emails that had arrived the night before. I then made myself a steaming hot mug of English tea and went to my Facebook page and checked my messages. One was from a woman I'll call "Alison Armstrong." As you'll soon see, there are certain, solid reasons why she wants her real name kept out of this book.

It's not an exaggeration to say that as I read Alison's message, the hairs on the back of my neck and on my arms rose up and a sudden chill came over me. I was about to have a conversation with someone…*whose dreams had been invaded by the Men in Black.* Yes, really.

I suddenly felt even colder. So much for the tea warming me up.

Alison wrote:

Hello Nick, I hope it's okay that I send this odd MIB story to you. In 1988 I had a prophetic dream involving MIB before I'd ever heard of them. Later that year or the next, I would read Whitley Strieber's *Communion,* and really wonder. Yet my dream had nothing to do with UFOs. This dream has bugged me for years. I have to be discreet and not give names or places as this story involves a crime - nothing I was involved with, yet was privy to.

The backdrop: I grew up in a small town in the Midwestern U.S. In our town was the one store for clothes and school essentials, like brownie/girl-scout/boy-scout uniforms/letterman jackets/gym clothes, etc. It was family-owned and the building was old, from the late 1800s, and part of the historic center of town. It was just past the main bridge that straddled the river. I had a huge crush on the grandson of the family who owned the store, and we were friends. I was 17 at this time, and he was a year older than me. His father had managed the store ever since I was a kid.

My friend graduated in the summer of 1988 and moved to the west coast right after graduation. We had very little contact after he moved, but he did send one letter, including his phone number, and I believe we had one phone conversation prior to this event. Then one night, in the late winter of '88, I had a dream.

In the dream, I was walking through town across the bridge towards my friend's family's store, on that side of the street. As I approached, an all black - and I mean all black...no shiny chrome or metal - 1920's type of car sped past and parked alongside the entrance to the store just in front of me. I got very scared. From out of the front seats of this weird old, boxy car, came two identical, tall, blond men in dark suits, which struck me as really odd, and from the back seat came a man dressed as a Victorian woman in full mourning gear...high-necked, full-length black dress, slightly bustled, hat, gloves, the whole thing. But this man was identical to the actor Zakes Mokae from the film *The Serpent and the Rainbow* - the really spooky voodoo guy in the film...but in this dream he had bright blue eyes.

I was paralyzed with fright when he looked at me. They went into the store and I knew something was very wrong - I knew they were going for my friend's dad, everything felt very terrifying, and they moved kinda weird...the one in the dress almost floated. It was like they were there / not there. They were very fast, and very silent. Then I looked up, and just ahead at the intersection (which is there in real life) was a banner going across the street, but it had the news going across it in real time, like an electronic billboard, and it said "So and so has been busted" (my friend's dad), while showing the face of another prominent man from town. I was full of utter dread but knew I had to go, and turned to run back the way I'd come, when suddenly the three people in black came out of the store and jumped in the weird car and started chasing me. I ran for about a block and then dove into a pile of snow, like the kind of snowdrift that gets left on street corners after the plows go by. I woke up, heart pounding. (The end of the dream would prove to be possibly precognitive too...)

After school on the following day, I received a surprise phone call from my friend on the west coast. I don't know why he chose to call me, but when I answered he just said, "My dad got busted last night." Turns out his father had sold cocaine to an undercover D.A. three times in a sting operation.

Now, I had been aware via my friend that his dad used drugs, but had no idea the extent of usage and certainly not that he was under surveillance. This dream has left me with so many questions. Why the Men in Black? Why the man dressed as a Victorian woman? Why looking like that actor? Why the 1920s vehicle? Why were they in a dream that was telling me of something I couldn't have known, yet the event predicted happened just about the time I was dreaming it?

Any insight would be amazing and welcomed. I hope you enjoy the story of the event, and thank you for all your work - it keeps the world a brighter and wider place!

Thank you.

I immediately wrote back to Alison:

> Many thanks for this. There are several things in your message that are very similar to things I have heard from other people. One being that many people dream about the MIB, but they are not like normal dreams. A lot of people felt that the MIB literally "invade" our dreams and manipulate them. I have other reports of prophetic dreams and the MIB. Also, the man disguised as a woman (in black) is very interesting. I have a new book out very soon titled *Women in Black.* I have 5 or 6 cases in the book where the where the Women in Black were suspected by the witnesses of being men in disguise, and some of them are from Victorian times, just like your account. I'll be pleased to mail you a copy of the book when it's published. All best, Nick.

It was all getting very weird. Soon, there was a follow-up message from Alison:

> Thank you. Do you know if there's anything historical around why the man as a woman in the dream would take the appearance of an actor who played a Voodoo person? As I recall, that movie had just come out at the time, and it frightened me quite a bit. So is the fear the point? But the freaky blue eyes! Paralyzing (I guess, like a zombie would feel.)
>
> I guess the real question remains as to whether these MIBs are tangibly real, or otherworldly energies. Are they directed by some other will, or self-directed? Because one of the most lingering senses from the dream is that weird "what the heck is this?" freaky feel of being totally out of anything known. I've had precognitive dreams since, but never seen them again. This one dream is the only time they've ever shown up for me. So why then? It's all so fascinating.
>
> I truly appreciate your input and interest. It's one of those experiences that never leads to any satisfactory conclusion.

I replied:

> I don't think the MIB or the Women in Black are human, I think they have paranormal origins (not alien). There are cases where they have turned up at peoples' homes and caused poltergeist activity, and they also threaten people who investigate things like Bigfoot and strange creatures. They are almost like psychic vampires, feeding on peoples' fears and energy. It's hard to know who is directing them

> but they clearly have an agenda to threaten people who dig into supernatural issues. It's hard to know why he had the actor's face, but if they can invade our dreams, they may have picked up his image from your subconscious.

Alison quickly got back to me:

> Yeah, that's about the best interpretation for that actor's form being used. And for any form being used in other-conscious realms. But that these ones were showing a government activity is interesting. It was a big thing in such a small town, however, as I understand it, the man who was arrested (now deceased) was really a low-level dealer who just happened to get caught by bigwigs, in hopes he'd spill the beans to the bigger fish, but he never did. So he served a lot of time. The whole point being that these W/MIBs were connected to government. Are they generally attached to social standards / well-being?
>
> My response: "There are some cases where they appear to be government types, but they also seem to have knowledge of the paranormal sides of all this. Maybe the government is working with paranormal phenomena in some way. That's a possibility."

It was a suitably strange way to start the day. Little did I know it at the time, but the following twenty-four hours would be even stranger.

6

"SHE DIDN'T SMILE"

February 27, 2016

"The WIB claimed she was a 'government attorney'"

The day began with me pondering on the idea of penning an article for *Mysterious Universe*, an Australian website I write for on a regular basis. The planned theme of the article was a curious one. I have in my files more than a dozen reports of people who claim encounters with MIB and Women in Black in two specific, semi-related locales: libraries and bookstores. There is, for example, the very weird saga of a man named Peter Rojcewicz who, in 1980, had a bizarre encounter with a Man in Black in the library of the University of Pennsylvania. It was an encounter with a tall, wizened old man with a European accent. As the aged and unsettlingly odd figure in black proceeded to grill Rojcewicz on the matter of flying saucers, Rojcewicz suddenly realized he was alone in the library with no-one but the creepy MIB around. *Everyone had vanished*; as in *gone*. Only when the old man exited the building did normality return. And then, so did the people.

On a brutally freezing cold, dark winter's afternoon in 1987, Bruce Lee – an editor with William Morrow – had a nerve-jangling encounter with both a WIB and a MIB in Womrath's bookstore on Lexington Avenue, New York; a store which just happened to have a lending-library. Lee was curiously drawn to the pair. No wonder, as they appeared to be speed-reading a book that Lee had personally worked on. It was Whitley Strieber's then-newly-published and soon-to-be-a-bestseller, *Communion,* which told of Strieber's alien abduction-style experiences. Lee described the pair as short in stature; their faces masked by sunglasses, hats and scarves. Eerie and unsettling giggles came from both the man and the woman, who exhibited clear hostility towards Lee, something which later led him to comment that the pair had "mad dog" appearances.

Ten years later on the other side of the pond, the wife of an English UFO researcher named Nigel Wright was threatened by a mysterious woman in white. It just so happens that on the very day Wright was busily pouring over old newspaper archives. Where? In his local library, that's where – and in search of material on UFOs in the county of Devon, which was where Nigel and his wife Sue were living at the time. Not only that, Nigel found in the archives an old report of what sounded like an encounter with a group of Men in Black in the early years of the 20th century.

Well, I thought, these and the various other reports in my files would make for a good, thought-provoking article. Before I even had the opportunity to start writing the article, however, something decidedly synchronistic occurred. To my utter amazement, the following message popped up at Facebook from a man named Mitchell Waak. He told me:

> I've met the MIB twice, 2005, about 2009, and a WIB once in the mid-1990s. The WIB claimed she was a "government attorney," had a strict intimidating manner. She had a black suit with dress, black shoes, no black hat, and a black briefcase. She was in a mall bookstore (Waldenbooks) shopping for UFO books, (strange). She told me she was looking for [Brad] Steiger's *Project Blue Book,* the last copy

> of which I bought day before. She followed me out of the store and talked to me briefly. I told her, "any ufologist (me) could be easily discredited by claiming they had watched too many *Star Wars* or *Star Trek* type movies." She didn't smile, and then we both separated.
>
> I've met about 5 different alien races since late 1950s starting with one group, The Elephant Skinned or Wrinkly Skinned Alien in about 1957-59 when I was 4-5 years old. This is the same group as per the 1973 case Hickson/Parker in Mississippi. Today, I am retired as a hospital pharmacist with Pharm.D. degree after 32 years of practice. I had attended UC Berkeley as science student back in the 1970s but did not graduate there. I have NOT been involved myself (not been involved in what? UFO research?) that much due to these MIB approaches and subtle telepathic and verbal threats. I am a private UFO/alien researcher.

The surreal nature of all this was hardly lost on me. When I was looking into the issue of "dream invasions" and MIB, in no time at all I had someone – Alison Armstrong – contact me about how the MIB had invaded *her* dreams. Now, within no more than mere *minutes* of planning to write an article on the MIB/WIB/library/bookstore connection, I received an email on that very issue from Mitchell Waak. Something was going on; something seriously odd. And I was tangled up in the black heart of it all.

I wasted little time in replying to Mitchell. A couple of days later we had a lengthy phone chat about his experiences with both extraterrestrials and the Woman in Black, who, he said - adding something that he had omitted in his messages – had extremely pale skin and a completely wrinkle-free face. Mitchell also shared with me, and generously let me use in this book, a copy of a letter he mailed to the late alien abduction researcher,

Budd Hopkins, back in 1992. It outlined Mitchell's early and then-still-ongoing experiences with non-human things:

> At about midafternoon in the fall of about 1958 (or 1959), a year before I was to enter school (kindergarten), I had an unusual and terrifying experience which I still can't account for to this day. My older brother was already in school and I was home with my mother and no one else. I was sleeping in mom's room during a mid-afternoon nap. My age must have been 4 years old. The house was located in back of an aluminum manufacturing plant in a small town in Wisconsin. I was old enough to use the restroom unattended, so I got up from the bed and started toward the hallway. The bathroom was midway down the hall on left. But as I approached the entryway to the hallway, I noticed a pair of strange feet at the opposite end.
>
> I slowly looked up and saw a dark gray, almost black human-like figure at the opposite end of the hallway. This figure had arms, legs, body, and head but had no distinguishable features (eyes, nose, mouth) and a dark grey very wrinkled type texture throughout the entire body, including the head and neck. This figure appeared tall and heavy to me (6-7 ft, and 200-300lbs), respectively.
>
> Naturally I was scared, and in that precise moment of my onset of fear, this "creature" told me somehow with an apparent mental telepathy, "Do not be scared, I will not harm you." Needless to say, I was scared to death by then and turned around and woke up my mother, who was sound asleep on the bed in her room. But, as I began to turn around, I noticed a slightly greenish but still transparent cloud with kind of white sparks had encircled me and I couldn't turn around all the way because I was

paralyzed. Immediately thereafter I lost consciousness. The next thing I remember is hiding and shaking with fear underneath my mother's bed, looking for the creature's feet to come into the room from my view underneath the bed. Apparently the creature left, and no further incidents occurred that I remember.

I believe I could have been abducted, but I don't claim to have seen any spacecraft or anything like that. I told this same story to my parents, brothers, people at work and school for years; I will always stick to my story. They tell me, "You had a bad dream," or "You saw a shadow, or you were trying to get attention." But how can a shadow talk to me without a mouth or paralyze me as I turn to run away?? And what happened to me after losing consciousness??

Today as an adult, this event is still very memorable 34 years later. I am still a very restless and "unsettled" person trying to understand what happened; I still want answers. This was a truly traumatic event, I remember very little about my childhood years, but always will remember this.

Why are the Women in Black and the Men in Black so drawn to bookstores and libraries? What motivates them to lurk among books both old and new? What was it that caused that undeniable shift in reality reported by Peter Rojcewicz as he sat in the library of the University of Pennsylvania and faced with an old MIB in 1980? These, and many more related questions, were on my mind when I got off the phone with Mitchell Waak.

7

"THE WITNESSES FELT THEMSELVES BEING CHOKED"

February 28, 2016

Creeps wearing gloves

On checking my email around 8:30 a.m., I saw one from a woman named Belinda, who had a traumatic dream two night earlier – she encountered a Man in Black wearing shiny black gloves. In the dream, the MIB pursued her along an old railroad track – one that was seemingly never-ending. Only when the MIB was within a few feet of her did Belinda wake up with an ear-piercing scream, something that Belinda's husband confirmed.

Over the years, a few reports have surfaced of the menacing Men in Black wearing black gloves, but certainly nowhere near the extent to which they are associated with black fedoras and old, 1950s-era black cars. Between October of 2015 and the early part of 2016, however, I received what I can only describe as a "cluster" of reports of the MIB wearing black gloves. In other words, Belinda's case was not alone. There were others, too.

Now, if I got – let's say – a report or two every couple of years of gloved MIB, I probably wouldn't have given the matter much thought. But, that was not the case: I received eight such accounts across a handful of months, not even including that of Belinda. And, from varying parts of the United States. Admittedly, at first, I thought this might have been some curious kind of MIB "spam" or "fake news." Namely, a group of people deciding to get together and waste my time with a bunch of bogus, concocted tales. But, having spoken on the phone with six of them, and having had extensive email exchanges with the others, I was quickly one hundred percent sure that this was not the case.

Of the group, six had experiences with the gloved MIB in what were classic states of hypnagogia. For those who may not be aware of the phenomenon, it is – as Wikipedia correctly notes – "the experience of the transitional state from wakefulness to sleep." It's a stage in which people can hear voices and unclear mumbling. Loud "bang-like" noises are not uncommon. Nor are the sounds of a door knocking or a doorbell ringing. Visual hallucinations can occur, too. Sometimes, such events are traumatic. On other occasions, however, they're uplifting. The phenomenon of "sleep paralysis" can also be a factor in hypnagogia, but certainly not always.

Now, the down-to-earth explanation for hypnagogia is that it's purely a product of the brain, provoked by that aforementioned hazy state of being semi-awake and partially asleep. Over the years, however, I have spoken with more than a few people who are absolutely sure in their minds that there is something else to all of this: A belief that paranormal entities can literally "invade" our dream-states. And, perhaps, our *semi*-dream states, too. Of course, there's no hard evidence for this, but that hasn't stopped such a theory from persisting. Particularly so in so-called "alien abduction" experiences, and also with regard to encounters with the likes of the "Old Hag," Incubus- and Succubus-type creatures. It seems that now we can now add the Men in Black gloves to that list.

As controversial as it may sound, I'm quite open to the idea that what many assume to be internal experiences may have an external "Fortean" aspect attached to them. One of the reasons being when I get these "cluster"-style reports. Now, I'm not saying that an entire bunch of people – spread all across the United States and unknown to each other – couldn't share weirdly similar experiences. But across just a few months? And all with the imagery of MIB wearing black gloves? And they all decide to contact me? That's when I began to think we're not dealing with just hypnagogia.

In all of these specific cases I received, the Men in Black were wearing hats. In two of the cases, the witnesses felt themselves being choked by the glove-wearing MIB – something that forced them to fight for their very lives, as they saw it. In three cases, the witnesses reported unpleasant odors in the bedroom at the time of the event. Importantly, in one case, the odor persisted long after the witness woke up.

Interesting? Yes. Intriguing? Definitely. Terrifying? For the witnesses, undoubtedly. For me? I saw it as yet further data suggesting more and more that the Men in Black phenomenon is one of paranormal proportions. "Secret agents" of the Will Smith and Tommy Lee Jones type? No way.

Then, there were the three MIBC cases that reached me. MIBC? That's a new term: Men in Black Capes. One such MIBC encounter occurred in 1969, in Pasadena, California. The second story came out of Albuquerque, New Mexico in 2002. As for the third encounter, that one occurred in 1997 on the island of Puerto Rico.

Despite the differences in locations, there were a number of notable similarities. All three cases involved women. All were in their twenties at the time of their encounters with the MIBC. Two had graphic memories of alien abduction-type experiences, and one had a still-baffling period of missing-time that continues to disturb her to this day. There were other similarities, too. Within days of their encounters, all three had strange and terrifying experiences in the dead of night. Yes, there were sleep-paralysis-style sensations, such as being unable to move, a sense of a hostile figure in the bedroom, and a feeling of profound dread.

What really stood out for me, however, were a couple of very specific similarities. For example, the black hats they wore were far more like cowboy hats in size and style than fedoras. All three believed the MIBC were there to warn them to stay silent on what it was they encountered. One of the "men" wore a thick, silver bracelet on his left wrist. Another wore a long, silver chain around its/his neck. And their cloaks were *very* long, as in down to their ankles. I'm used to getting amazing and strange reports – of all manner of things – on a regular basis. But, to get three such accounts spanning more than thirty years from distinctly different locations, and in which the circumstances were practically identical, was noteworthy. It still is noteworthy. I also find it curious that I should have been on the receiving end of all three cases.

I still get such cluster-based reports today. They remain as thought provoking as they are eerie.

8

"YOU KNOW WHEN THEY'RE AROUND"

March 17, 2016

"The color of their skin was wrong and their eyes looked dead"

A fascinating story came to me on this day from a man named Brian, living near Conroe Lake, Texas. Over the course of several emails in early 2016, he shared with me the extensive details of a veritable wave of hostile encounters that had turned his life and that of his family almost upside down. Brian's story was among the most disturbing of the most disturbing. Why? Because they involve people who not only confronted by both the MIB and the WIB, but who fell seriously ill after their encounters – almost as if they have been supernaturally infected. Which may actually be the case, as unsettlingly as it sounds. Brian told me:

> OK, I'm just going to lay this out as it happened. About 7-8 years ago I was a MUFON member, working on becoming an investigator and started looking into some local cases, performing video analysis, etc. The one thing I got out of this experience was that old John Keel quote – "the more you look into this phenomenon the more it looks back."
>
> Now, my interest in Ufology began as a child, my Mom had some experiences (asking her for details always

resulted in her kind of "forgetting" the conversation, it was just impossible to get more than cursory info from her) and being an early reader with a library at home that included *Chariots of the Gods* and other related books. As a kid in the 70's Ufology and cryptozoology were a lot more "accepted" in popular culture.

Anyway, trying to keep this moving for you - as I got more involved talking to witnesses from the Stephenville [Texas UFO] incident [of January 2008] I started to have a lot strange issues. Weird beeping noises, some degree of paranoia, movement, sounds in places of the house where no one should be, etc. I started sleeping in weird places. Later I theorized that I was trying to put myself in between areas of ingress and my wife / kids. The floor, the stairs, etc. I would just wake up in strange places and have no clue how I got there.

So, on this day I woke up and I remember being in pain, I couldn't breathe and my upper body was collapsed forwards and sideways. I saw a concrete floor, a lot of foam below my head and realized I was in my garage, in a lawn chair and was facing out towards an open garage door. What woke me, and what I continued to hear was my wife calling my name and her pitch and intensity rising. My wife is a very, very cool customer type. She doesn't get excited at all and I'd never heard her voice sound panicked like this. At its peak, she had come through the house door to the garage I was waking up in and was maybe 10 feet to my right and behind me.

At the same time I heard her slap the button to close the garage door and as I raised my head I saw two pairs of black shoes, then what became a man and woman dressed in black dress clothes. I thought "realtors" first, but as I got to their upper bodies I kind of froze for a second. They were both leaning over me and their faces just looked, weird. The color of their skin was wrong and their eyes looked dead. It scared the hell out of me but the garage door slid down in between us and my wife was there now, pulling me up and into the house. She asked me "what the hell were those?"

Our front door was close, a left angle turn from the garage and my flight response had turned to fight at that point so I went straight to the door and out and should have been able to see the couple to my left but they were gone. We lived on a cul-de-sac at that time, there was nowhere they could have gone that I wouldn't have been able to see them. I went out, circled the house - nothing. All I saw was what looked like a pair of drag marks on the driveway going into the garage - which really freaked me out. I hadn't really come across any MIB instances with a female involved so I didn't really connect it until I started to hear about it, namely from your interviews and put it into the context of everything else that was going on at that time.

My wife is NOT into Ufology, her Dad's a preacher - she thinks anything paranormal is caused by demons or evil spirits or whatever. All I could get out of her on this incident was that she saw the same thing I did - the couple standing over me and that it all looked / felt wrong, they looked all wrong - eyes, skin everything I saw in my quick look. I was really sick after this for a week also. No one else in the house got sick at that time and I had some really weird physical symptoms including a large and sudden drop in my testosterone level which normalized after. (Only found out because my Doc was trying to work out why I was so fatigued.) I don't know what to make of that but it occurs to me "if" I had some kind of incident that other people with the same thing might not be tested as extensively as I was at that time. It might be unrelated - just thought it was really strange combined with the other symptoms.

We had some other issues after this but that's probably more than enough for one email, I don't know if it helps your research but my investigator side says every data point contributes so here you go. My oldest daughter was impacted to this day, even our neighbors at the time were drawn into the events. I haven't shared this story with anyone in the field, just a short post on ATS [Above Top Secret] once a long, long time ago.

I love your work, keep it up. I'm an MU [Mysterious Universe] subscriber and love being able to read your stuff.

Thanks, Brian.

Later that same day, and after I replied to Brian, he sent this to me:

There was one other incident but it was a bit different. I had started working with a reporter, connecting them to a source I had and around that time this black SUV kept showing up outside my house. I noticed it but I didn't really think anything of it, thought it was someone visiting a neighbor type thing. Then I noticed there were people in the SUV and felt like they were watching my house. I thought I was going full bore paranoid at that point so I finally walked outside, walked past the truck to get a closer look and maybe start a conversation. As I walked up, the window came down about 3 inches and the person in the driver's seat called out my name. Then the window rolled up, the SUV started and they drove off. The really weird thing was I just "knew"" I wasn't supposed to push forward with my plans w/ the reporter. I don't know how or why but once I made that decision the visits stopped immediately.

There was some other stuff around that time, One morning, my neighbor started ringing my doorbell like a madman at 5 a.m., He kept pushing it non-stop a few dozen times. We were all asleep so by the time I got to the door, he was gone. He'd left this note that the "noise from our house was keeping them awake." There was no noise that I could detect, we were really puzzled over it and when I talked to him later about it he insisted there was this low frequency hum type noise coming from the direction of our house that was shaking them out of their beds. We heard absolutely nothing. He was very firm that it was coming from our house and that it didn't stop until he was at our door. They moved out very quickly after that.

My oldest daughter, 16, still sleeps with a light on. Something scared her badly one night but she's never been able to describe the issue in any detail. She began having some bad anxiety issues around that same time.

Other than that, during that time I did have a couple of sightings but they were distant. One was a black object that was looping around in the clouds, very, very quickly. It just didn't look "right." It didn't move like a plane or a bird. The other time was just a very bright light at night in the sky above me while I was out walking. The really weird thing is that it seemed to react to my questions, I'd think, "is that a UFO?" and it would move from side to side. "I wonder if they can hear me?" and it would move again. Otherwise, unless I was asking a question it just stayed perfectly still. The first one I can remember pretty clearly, the second one I have a harder time with.

To be honest, if there's more to my own incidents I don't want to know. The few things I've read lead me to believe people who pursue these kinds of issues, looking for repressed memories and crap like that only end up worse off for it. I'm still interested in the phenomenon, can't shake that but once it began to touch the people around me it became a lot more menacing? Not sure of the best word to use there.

There was one other incident with a MIB type when I was a child w/ my Mom but I need to think about it. I was really young and things are really fuzzy on some things while being very clear on others, like they happened yesterday. It was really bizarre so I need to reflect on it more and see if I can explain it with any kind of clarity.

And things were not quite over: Brian sent me a third email.

Nick: Just wanted to point out that was a really weird time for us. I know how crazy some of it sounds and want to assure you I'm a normal, boring person who does not have paranormal experiences on a constant basis. We're mostly happy and just raising our family in this nice enclave on the lake now. I find the subject fascinating but it's almost not a healthy thing, more of a WTF happened to me / us thing? I hope you understand what I mean. I don't tell anyone these things, can't even discuss it with my wife, who I love dearly and consider my best friend. I'm not looking for attention, for some reason I've always identified with your work and feel I share a common outlook in some ways based on your writing persona / attitude.

Hell, I'd laugh at some of the things that have happened to me if a witness were relating them to me, other than

I've seen how truly strange things get and like you, have heard some things often enough to bypass coincidence as a consideration. Seeing things first hand really adjusts your outlook, so to speak.

One thing I forgot to mention, I don't know if you've ever eaten anything that's made you really sick and then afterwards, just the thought or smell or even sight of that food will bring back the symptoms again? That's how it is with these MIB things, pretty much exactly. You know when they're around. You can feel it like the oncoming symptoms of a flu or food poisoning attack.

I'm glad to be able to share this stuff with someone, I do thank you for taking the time to "hear" me. Thanks, Brian.

Brian's words echoed those of a source I mentioned in my *Women in Black* book and who asked me to identify him only as "A Hesitant Believer." AHB told me of a fear-filled encounter with an MIB and a WIB in Tampa, Florida in November 2008. In closing, he said to me: "I have a question for you: In your studies have you ever come across individuals that came down with severe illness after encountering these 'people?' Only three days after I saw them I developed a severe bacterial infection on my legs and lower abdomen. It was a Strep infection, and the doctor could not determine how I contracted it."

Quite clearly, there is something undeniably dangerous about the MIB and their various offshoots.

9

"HIS FACE WAS SO THIN AND HIS SKIN WAS STRANGE"

March 17

"All three wore the 'Frank Sinatra Fedora hats'"

March 17 was, to say the least, a busy one when it came to the matter of the MIB. As well as receiving Brian's account (detailed in the chapter immediately above), I also got a lengthy MIB-themed story from a man named Kent Senter – who I got to meet in late 2016, as will become apparent later in the book. I'll present Kent's story in uninterrupted fashion:

> I wanted to share some experiences with you and, as I stated, have an incurable cancer I have been fighting. First, I have had a few UFO/UAP encounters that go back to the 60's. I will try to summarize as my sightings would take too long to explain. I grew up outside of Washington D.C. as my father was employed by the government. When I was around 10 years old, I witnessed a disc that illuminated a beam of light that stopped before reaching the ground. I was sleeping/camping outside the back of our house when this occurred around 2:00 am - a bright orange glowing disc that was completely silent and still until if glided off. This aroused my curiosity.

My best friend's father was a deputy director for the FAA and was a retired colonel in the Air Force who flew under Curtis Lemay during WWII. After my sighting, I inquired from him if UFOs were real and he looked at me strangely then told me that he knew someone who walked into a hanger and saw a recovered disc and they locked him up in a mental illness hospital. Asking him why they did that, he stated that when the guy got out and if he talked, who would believe him? As everyone was very patriotic at the time – especially around D.C. during Kennedy's Camelot years – I was shocked and for a young kid, it scared me to death and I never talked about what I saw. That was experience one.

Recently, this past year I read an article in the *MUFON Journal* about a young kid's encounter with EBE's and a craft sighting that occurred less than 2 miles from where I lived, happened around the same time, as best I can tell. I showed my wife who was shocked as she is familiar with all my experiences.

Next, these strange coincidences that I will share all happened around the month before I held my conference in June 2013. I was asked by Leslie Kean if I would do an interview by Lee Speigel with the *Huffington Post.* After agreeing, I spoke with Lee and he asked me to tell him about my experiences and why I was sponsoring a conference with so many prominent speakers. I told him that back in 1975, when I was 21 years old and living in Durham NC, I saw a news report on TV about some UFO sightings in Lumberton N.C. I immediately drove down there and rode around for a while when I came around a bend to a long stretch of road surrounded by farming fields. I noticed a police car pulled over and some people standing in a field, looking around. I witnessed a red, faintly glowing object above the trees and off in the distance past the field. After looking at the object, I drove off chasing it but lost it visually during my pursuit.

Lee stopped me at this point and asked me if I knew about him and his experiences. I stated no. He then proceeded to tell me that it was he in the field back in 1975 and that I arrived at the tail end of an unbelievable experience. He was there investigating for J. A. Hynek and CUFOS. He also stated that he was the organizer of the U.N. UFO hearings with Gordon Cooper. Hynek, Vallee and I actually have a picture of that group that I have kept for years... and sure enough, the *Serpico*-looking gentleman in the picture is Lee Speigel himself. What a coincidence that I saw him 41years ago and this is how we met. We have had the pleasure of having dinner with him at the MUFON Conference in 2014 and he is quite the gentleman.

I also related about a sighting I had in the mid-80s in Durham N.C. I had a film crew recreate this sighting for presentation at the conference and it is on the DVDs of our conference. Anyway, this sighting was the final straw that determined my desire and intentions to research and find out what I had seen, as this was definitely some type of craft and not glowing lights. I located a N.C. UFO researcher whose name was George Fawcett. After meeting with him, we organized a steering committee and started the N.C. chapter for MUFON in the late 80's. This is when I officially became an investigator for MUFON. Nothing astonishing or earth moving until May 1993.

I was attending a presentation by Timothy Good in Raleigh N.C. when I received a call from our state director, asking me to contact the Alamance Co. Sheriff's dept. to help with a sighting and cattle abduction case with a 70 year old farmer. After interviewing the witness, I took a rookie investigator to the area where the cow was last seen - inside the edge of the woods next to the field. There was a gaping hole in the upper foliage and you could see blue sky, with branches broken upward. I filmed and tried to zoom in to see if there was any evidence of cowhide high up on the broken branches. I drove back to my residence after finishing.

I worked as an apartment and mobile-home park-manager at the time, before becoming a regional VP of operations for a large company. My residence was also the office for the park. From time to time I would get police / detective visits to review some of the residents' files that were kept under lock and key. When the MIB's came to the office door, I thought it was a law enforcement visit, just two days after my meeting with the Deputy Sheriffs and farmer.

Initially, I want to share with you that I have described this incident to my wife when we first started dating in 2003. I wanted to be open with her about my interest and experiences in case our relationship became serious, as there is such a great taboo concerning this phenomena. I wanted to be honest about my interest while finding out her feelings about the topic. Patty has never paid much attention in her life to the subject and basically knew nothing about it. I have also been on record with the film crew, as I wanted my experiences documented for the conference.

Through my peripheral vision, I noticed three *men* approaching my office door. I quickly glanced up from my work and immediately started to unlock the files on my right side, as it was in my mind that detectives were coming to view a file for some resident. After unlocking the files, I looked up and immediately felt numbness and somewhat of a tingling electric shock in my stomach. I have never felt this sensation before or since. It was similar to a funny bone hit, but in the stomach only. It was very strange, but as this happened I was looking at something that I knew was not normal.

From the inside, the front wooden door to the office was open and it opened inside and to the left with the inside doorknob on the right. Sitting in the living room/ office, my desk faced the wall and window just to the right of the front door. The glass storm door was closed, which they opened and came through. The first being walked in and had to bend his head over to get in the door. His face was thin and his skin was strange and almost a see-through milky color. There was no hair that I could see.

Next a short stocky *man* came in and then the third *man* came in, but when he did, I knew something was wrong and terribly strange as the third man was an identical twin to the first tall one. Both were the same height, same thin face and milky complexion and hairless. All three wore the "Frank Sinatra Fedora hats" and all had on sunglasses and suits. The third one went around the short stocky one who stayed to the left of my desk and was the only one who spoke. I tried to see the third one who went behind me and to my left but for some reason, I could not turn around to see him and only saw him through my peripheral vision. I struggled to understand why I couldn't turn around. It seemed he was behind me to the left and his hands were bent over as if his fingers were trying to touch his wrist below his palms.

It all seemed so strange and I felt like I was in a stupor. I immediately looked at the first tall one, who was standing with his back against the front wooden door and was looking straight across the room and not toward me. This is when I knew something was wrong, as the short stocky one was talking loudly and I couldn't hear or understand him, as I was mesmerized with the first tall one. Then at the exact moment I thought, "I want to see your eyes" in my mind. The first one immediately dipped his chin downward so that I could not look under his sunglasses. I then felt he was reading my mind, as if I had said it out loud. I kept saying in my head that I knew he could hear me and I kept repeating that I wanted to see his eyes and to take off the glasses. I don't know how long this lasted, but it must have upset the short stocky one, as he moved more to his left in front of the first one and started screaming for me not to go back out to that site, and yelled "Do you understand?" All the while he ripped his sunglasses off, as if he was upset with me for thinking this.

Nick, his eyes were almost perfect circles and the brightest blue with large black pupils. No eyebrows or hair. They left after I mumbled I understood but I still could not move for another minute. I got up to get the license plate number but all I could see was that it was government plates. The car looked old. After being back in the office for a few minutes, I realized that my VHS camera was opened and the tape removed. This camera was in the kitchen on the counter to my right and I don't remember them being on that side of the office. That was the end of the encounter.

This brings me to the final two coincidences. A childhood friend of mine from Northern Virginia whom I grew up with eventually moved here in Burlington N.C.,

Alamance County. He became a policeman and retired as a detective. I told my wife before the conference that I was going to ask Wayne over to get his help with his connections so that I can try to find out how the case was resolved with the sheriff's department and if anyone who was with me that day might have also received a visit .

I never gave anyone my name or address or anything. I introduced myself as an investigator with MUFON who was responding to their call, and we were whisked to the site before I could finish my introduction. I never understood how these *men* had my name and where I lived and worked. Anyways, Patty and I tried to think of a way to break the ice about my UFO interest and experience and I started by telling Wayne about my sighting where we grew up... he immediately interrupted me and said: "Yeah, I know; don't you remember the big rectangular craft we saw that night that was over our heads?" Patty and I were shocked as I don't remember a sighting with Wayne, and in my sighting, I was alone with a different shaped craft.

Another point is that Wayne is a no-nonsense guy and we have never ever had a discussion about UFO's in our lives. He knew nothing about my experiences as I told no one, not even my family. I was young and naive, and scared to death that something would happen to me if I spoke about what I saw.

The final coincidence was the video of the MIBs from the hotel in Niagara Falls. I saw this for the first time the month before the conference in 2013. I was stunned, as was Patty, because the girl's description of the MIB's and their actions are identical to what I told Patty in 2003 that I experienced in 1993. Nick, I know you research this and I wanted you to have a record of these events as I won't be around forever. I want to tell you that the event shook me up enough for me to stop investigating for MUFON and back out of this whole topic. I did not keep up with any news on the subject for years and years; until I became sick. I will admit, I went to the next MUFON meeting and when asked about the case, I blew it off as devil worshippers, etc. and that there was nothing to it. Don't know why I did that!

The State director who called me was George Lund but I could not find any information from him or MUFON to determine who called from the Sheriff's Office and I don't remember their names, but I did get the exact date the call came, from Timothy Good, as he kept excellent records. As I said, I was present at Good's talk when the call came. I have made four attempts to speak with someone over the last two years at the sheriff's department, but my contact has always been busy, or I have been very sick. I have been on chemo pills for a year-and-a-half now. I am going to try and follow up as I have always been interested to know if the three deputies ever received a visit. It seems I may have met the same ones as in New York, but maybe they are all similar. I also wanted to let you know that Patty bought your book on MIB's and the picture in the book of an MIB look like the same stocky man I saw. All it takes is one look and you would know something wasn't right with them...the stocky one was the most human of the three, but he and the other two were definitely not from here. I thank you for your time and appreciate you letting me share this with you. I also have another connection you would be really interested in as, historically, he was in very high places with this phenomena.

10

"THE IMPULSE TO BURN THE BOOK IS MIGHTY INTERESTING"

March 24, 2016

"No-one's gonna believe this"

On this day, Dennis Carroll, a friend who has a deep interest in the MIB controversy, told me:

> About a year ago, I was a guest on Lon Strickler's Radio show *Arcane Radio* and I kind of jokingly issued an invitation to the Men in Black. I told them to come and see me. Well, I kind of forgot about that statement and a couple a months went by and one evening I stopped at a nearby convenience store. As I came of the store, I noticed what looked like an almost brand new looking black Cadillac a good way out in the parking lot. It looked like a '63 or '64 Caddy. About the time I noticed it, a greenish kind of light came on in the interior and I swear I saw two men in black suits intently looking my way. They were also wearing aviator sunglasses. At night at that. I couldn't really believe what I was seeing. So, I reached for my cell phone to take a photo as I thought that no-one's gonna believe this. I discovered, however, that I left my cellphone in my truck.
>
> At the same moment, the green light went out in the car, the Caddy started slowly easing out of where it was

> parked. I rushed to get in my truck with the intention of following these guys to get a picture. By the time I got quickly rolling, the car was completely, almost supernaturally-like, gone. And that place has a huge parking lot. There is no way this vehicle could have gotten gone that quickly. I think I may have just gotten an answer to my invitation. Thinking back, I feel like that they definitely wanted me to see them. They were sending me a message of sorts. I have written several books and researched and investigated the paranormal for over forty years and that, other than the very dramatic UFO sighting I had several years ago, is one of the strangest things I've had happen. But ever since that sighting a lot of weird things have taken place.

Yep, "weird things" and the MIB are definitive bedfellows.

March 29, 2016

A legend passes

I knew that, in all likelihood, it would happen soon. After all, he was getting perilously close to a century in age. But I have to say, it was still unfortunate to see the news when it finally surfaced. You are most likely wondering what, and who, I'm talking about. It's none other than the passing of Albert Bender. He was the guy who began the Men in Black mystery back in the early 1950s, and whose 1962 book on the subject, *Flying Saucers and the Three Men,* remains a must-read for MIB enthusiasts. Despite having been threatened, taunted and tormented by a trio of blazing-eyed Men in Black in his Bridgeport, Connecticut home all those years ago, Bender lived to the ripe old age of ninety-four – proof that being confronted by the MIB does not necessarily mean death is looming just around the corner!

Although April 14 was the date on which the story of Bender's death surfaced on the Internet, it turned out that he had passed away a couple of weeks earlier, specifically on March 29. Both Loren Coleman and I wrote obituaries on Bender, which got sizeable coverage and commentary online. In part, mine read like this:

"Given that there have been claims Bender died in the early 2000s, a few people have already asked me if the Wiki page is accurate. Yes, it is. Bender had a full life and outlived just about everyone else who got involved in Ufology in the late 1940s and early 1950s. Over the years I have written quite extensively on the life and experiences of Albert Bender (my book, *Women in Black: The Creepy Companions of the Mysterious M.I.B.*, delves into his experiences even more). And *Flying Saucers and the Three Men* and *They Knew Too Much about Flying Saucers* – both Bender-themed - were books I read as a kid. As many people know, the mystery of the MIB is one of my particular interests. It's likely that interest would not have developed as it has without the words and experiences of Albert Bender.

R.I.P., Mr. B."

And, then, things took an extremely weird turn.

March 31, 2016

Another passing and a Men in Black connection

March 31, 2016 was the day on which the death of yet another ufological old-timer, Trevor James Constable, was revealed by Loren Coleman. Albert Bender had passed on just two days earlier. It so transpires that Constable had a deep interest in the Bender / MIB affair and concluded, back in 1962, that Bender's MIB were definitively occult-based in nature and origin. How do I know this? Because back in 1962, UFO writer Gray Barker published a follow-up, and far less known, title to Albert Bender's *Flying Saucers and the Three Men.* It went by the moniker of *Bender Mystery Confirmed.* It was, basically, a 100-page collection of letters from readers of Bender's book and who wanted to offer their thoughts and theories on its contents. One of those people was Trevor James Constable.

The author of *They Live in the Sky,* and someone who believed that at least some UFOs are living, jellyfish-like creatures, Constable wrote the following letter to Gray Barker, which the latter duly published in *Bender Mystery Confirmed:*

Dear Gray,

It is difficult indeed for me, as an occultist with some first-hand experience of this field of UFOs, to sort out Bender's journeys back and forth across the threshold line between the physical and the astral. A biometric examination of Al Bender would probably indicate similar things to what it revealed about certain other researchers - total inability to distinguish between events on two planes of reality.

Bender's honesty I do not for a moment doubt. His discrimination I would deem non-existent. It seems almost incredible that the man could relate the full story of the construction of his chamber of horrors in the attic in the way Bender has. This is what convinces me of his honesty. Nothing could be more logical, in an occult way, than that the invisible entities he invited by the preparation of this locale, should indeed manifest to him, and thereafter proceed to obsess him for a protracted period, using hypnotic techniques that brought the man completely under their control."

As to the nature of the entities involved, it seems that my writings about the "imperceptible physical" as source of many space ships, or so-called space ships, are only too close to the truth. Indeed, if Bender's experience has any value, I'd like to suggest that it certainly illuminates a re-reading of *They Live in the Sky*. I don't believe I know of any case quite like Bender's, where a man seemingly oblivious to the reality and laws of the occult, brought upon himself the energetic attention of aggressive occult forces. Certainly, the man can thank some kind of Divine intervention for the preservation of his sanity - if everything he writes is true.

Assuming that Bender has been truthful and honest, I would say that the lesson in his experiences is this: For the understanding of the UFOs and all the bewildering phenomena connected in this field, a working knowledge of occult science is indispensable. This lesson, driven home in innumerable ways since saucers came to mankind, is given new force with the Bender book. But few there will be who will heed it.

It does not surprise me to learn of the various manifestations you report – in fact, an occultist would be surprised if they did not occur. The psychic lady in Cleveland undoubtedly provides enough prana ["prana" being a Sanskrit word meaning "life-force"], voluntarily or involuntarily, to permit the near manifestations of low grade entities drawn to her aura by her concentration on the book. The odor of drains and these occurrences in the toilet are old hat to occult students. It might give you pause to wonder just

> what you are setting foot on and undoubtedly drawing to yourself. Not for the sake of money did I suggest that you leave this kind of thing behind! The impulse to burn the book is mighty interesting. This impulse springs from the best kind of intuition. When the Fire Gods come, everyone else in the unseen departs. It's as simple as that, so the lady in question would have been well advised to consume the book in fire, as her intuition so rightly prompted her. The lady who returned the book was also wise, if she felt that way. Send it back to its source – a sound occult system of personal defense, if no higher knowledge is at the individual's disposal.

What Constable was talking about here – with regard to one reader burning her copy of Albert Bender's *Flying Saucers and the Three Men* and another one who felt so unclean and infected after handling the book that she felt the need to mail it back to Barker – were other people who had written to Gray Barker and whose accounts Barker privately shared with Constable.

When I dug out the old copies in my possession of the correspondence between Barker and Constable – specifically for inclusion in an online article on the pair – the latter's words about the woman's decision to "burn the book" instantly reminded me of the words in my Introduction to *The Black Diary*, parts of which, in draft-form, were written around eight months before Constable died. I wrote, you'll recall: "...if I were to tell you to burn this book after reading it, I would only be half-joking. In some of my darker moments, there's absolutely no joking about it, at all."

I have to admit that when I read about the burning of the Bender book, I couldn't help but think there was "something in the air," something that led me to make a similar comment. I hadn't delved into the Barker-Constable letters for at least fifteen years, so it's unlikely I was inspired by them to make a similar comment. For me, it was yet further evidence that some books are dangerous in the extreme. And, if one has a certain degree of insight, one can pick up on such things. Bender's *Flying Saucers and the Three Men* being one of them. Maybe, too, the very book you're reading right now.

11

"I KNEW YOU WOULD COME"

April 15, 2016

"Mr. Atol had been a perfect example of such weirdness"

Just one day after it was revealed Albert Bender had died on March 29, I received the following account from an Englishman named Gerry Banyard, now living in the United States. It was without a doubt one of the strangest of all MIB-based cases I have ever come across.

> The experience first began one Saturday teatime outside a Manchester [England] bar in the summer of 1994. It was around 80 degrees. I sat alone enjoying a drink when I noticed a man around 35 at the table next to me talking to two girls. He was strangely attired in a black roll-neck sweater and, even more bizarre, black-leather trousers on such a hot day. He was doing all the chatting and, most noticeable of all, laughed in strange fashion, almost uncontrollably at times. He looked over and invited me to join the three of them. I declined at first, but accepted his second invite. I soon learned that they too had just met him and it was not long before he offended them by making some inappropriate comment that I did not fully hear. They left somewhat abruptly and I was left all alone with him.

He introduced himself and – like something from a [Harold] Pinter-play - announced that he was "selling gold," and should he disappear then, it was because he was meeting a prospective buyer. I noted that he clutched a bag that he did not seem to wish to let go of. His voice was high-pitched and, true to word, he said, "I'll be back soon." Bag-in-hand, I assumed that it was full of the gold stuff, but thought it strange that he should be selling such at this time on a Saturday. His name was Edward Atol, and, it was not long before he returned announcing that "the other guy hadn't showed." I thought: surprise, surprise. He then sat next to me asking "where" he could "buy bread" in Manchester. And, he was not joking. I said that there was a Tesco-Metro but he opted for a close-by newsagent. He returned with a "milk-roll" and opened up his black bag revealing nothing more than a dictation-machine and a pack of cuppa-soup. He proceeded to offer me one of them, but I declined his offer.

At this point I thought that maybe I had stumbled on one of those "classic" Men in Black-type entities. Or, that he may have found me. But, not by chance, as I was very much into UFOs and such phenomena in the early nineties. This was "big-time" Ufology. I went along with his unusual banter and mannerisms. I thought, "Okay, I'm watching you as much as much as you [are] me. But the difference is you don't know it."

I spent a couple of hours listening to his idiosyncrasies and strange sayings. I told him that I was about to leave, as I had to rise early the next morning for two weeks in Mallorca. He said that when I return we should meet again. We agreed to do so, some two weeks later.

I had noticed many strange things about Mr. Atol during our two-hour encounter, and one was that he did not perspire, although wearing leather pants. I noted such and, on the way home, I popped in to see ufologist, Arthur Tomlinson. He said, "Play along with him and find out what you can."

I returned from sunny Mallorca and wondered if Mr. Atol would keep his promise. To my surprise he was there, still clutching his black bag. He opened it, saying to me, "I knew you would come." He pulled out the small dictation machine, which I was curious about. And, having noted when he opened his wallet that there were neither no notes nor cards of any description, I offered to buy it. We somehow agreed on a price of fourteen pounds. I spent around three hours with him. Once again, going along with his strange sayings and pretending to laugh when cued, but not really understanding what he had said.

A couple of years later, I read John Keel's *The Mothman Prophecies* and noted that the author spoke of MIBs who spoke of gold, and who would not necessarily show any interest in UFOs. They had bizarre laughter and sayings. Mr. Atol had been a perfect example of such weirdness. Having arranged to meet Mr. Atol the following Wednesday, I once again called on Arthur. We listened to the tape and amongst such ramblings as "leather settees' and "coffee-machines," we could not identify the sense nor, at times, the language he spoke. Nor did the rhythmic, annoying metallic sound continually in the background help us.

I met Mr. Atol for the final time, as arranged. The only change in his general banter and weird mannerisms was his ability to use poetic terminology; as though testing me, in order to gauge a response. His final strange act was to leave his wallet on the bar before heading for the toilet – again as customary practice. I could not help but peer into it. Not surprisingly, there were no bank notes, cards, nor ID of any description. Needless to say, Edward Atol vanished as mysteriously as he had arrived, leaving me none the wiser for my experience.

Atol vs. Apol

There ended Gerry Banyard's very strange story. I knew, of course, that Gerry may have simply crossed paths with an eccentric character of the type we all run into now and again. There were, however, a number of stand out issues that led me to believe that was not the case. First, there was the matter of the name of Gerry's strange man: Edward Atol. His surname, Atol – or, perhaps, the surname he chose to use when speaking to Gerry – was very similar to Apol. The latter just happened to be the surname of an equally bizarre figure who turned up in a 1967 MIB/WIB case investigated by John Keel.

Jaye P. Paro was a woman who had a number of bizarre, UFO-themed encounters in May 1967, in upstate New York. She was also a host on Babylon, New York's WBAB station. On one particular day, Jaye decided to take a walk. It was barely dawn, and the town in which she lived was still shrouded in shadows. As she walked passed a particularly dark alley, a Woman in Black loomed into view, as if from nowhere, or from some nightmarish realm.

Then, out of the blue, came a black Cadillac, the absolute calling card of the MIB. It came to a screeching halt next to the two women, and out of one of the rear doors came an unsettling-looking character. It was a man dressed in a dark grey suit, with an "oriental" appearance, and who sported a disturbing, almost maniacal, grin. The driver seemed almost identical in appearance. The man with the fearsome grin shook Jaye's hand

and said, "I am Apol." Jaye said that holding Apol's hand was like holding the hand of a cold corpse.

The man gave Jaye a piece of parchment that contained a metallic disc, in terms of size around that of a quarter. Jaye - who later said that throughout this entire, odd experience, she felt light-headed and spaced-out – announced to the WIB and Apol that she was going to mail both items to someone, who happened to be John Keel, although she did not tell the pair who, exactly.

Evidently, both the WIB and the MIB were happy, since they again suddenly sported cold, eerie grins. The black Cadillac immediately returned and whisked the pair away for destinations unknown, but surely no good.

According to Keel, when the package arrived it contained nothing unusual; just an old envelope and a nameless ID tag. Clearly, this was not what Jaye had mailed to Keel just days earlier. So, Keel mailed the items back to Jaye. Cue even more strangeness: When Jaye opened the package, she was overwhelmed by a smell of sulfur – a classic aspect of paranormal and demonic activity and presence, and something which Albert Bender noted in his experiences with the Men in Black in the early 1950s. On top of that, the disk was discolored. Originally silver it was now a very appropriate black. Keel said of this situation: "The implication was clear. Someone had the ability to intercept the US mails and tamper with things in sealed envelopes!"

The bizarre activity continued: Jaye had further, nonsensical run-ins with Apol. She was visited and quizzed by mysterious, military men who turned out not to be military men, after all.

I couldn't fail to note that the similarly named Atol and Apol both had maniacal grins. There's also the fact that Atol had a thing about gold. As I knew very well, there are a number of cases on record in which the Men in Black displayed deep knowledge of alchemy – the secret, occult practice of turning base-metals (such as tin and lead) into precious metals (like silver and gold). Weirder still, at the time Gerry sent me the details of his story, my *Women in Black* book was about to go to print. It contains a chapter on nothing less than the WIB / MIB connections to gold. And, in the very same book, I told the full story of Jane's creepy encounter with Mr. Apol.

The issue of gold, the similarities between Apol and Atol, the unsettling grins, and the synchronistic fact that my then-still-forthcoming *Women in Black* book contained the very elements present in Gerry's email, led me to conclude that, yet again, outside forces of a supernatural and manipulative nature were stirring the paranormal pot, and for reasons that only they truly ever understood.

12

"THE WOMEN TALKED WITH A LOW KEY MONOTONE SOFT VOICE"

May 25, 2016

"The women were unusually pale"

From Paul Don Roberts, I received this:

> In an interview with Richard Rivas, Richard makes claim that many UFOs have been seen on his property. Some of these UFOs have actually landed on his property. Richard made claim that he actually saw Gray aliens in and around his bushes and trees. When Richard's story about this ranch was finally picked up by the Biography Channel's *My Ghost Story*, it was 2 days after the filming that he was visited by two women. The two women were dressed in black clothing attire. Both wore skirts. Richard said that both women looked appealing and both women sort of looked like each other. One of the women claimed to be a "consultant for NASA" and asked about the paranormal activity at the ranch, especially the UFO sighting. Richard let both women into the house. The women talked with a low-key monotone soft voice. They asked him things like "how long have you lived here;" "Describe the aerial phenomenon;" "Did they land on your property;"; "Did they give you a message;" "Why do you think they gravitate to this

property;" "Who have you told this story to;" "This is now under investigation and we expect your confidentiality;" "You do understand that if you disclose any information, that you could do Federal prison time?" During the interview, Richard noticed that the women were unusually pale and their eyes reminded him of a cartoon anime character; they were somewhat large and oval in shape. When they walked out of the house, they seemed to walk in unison. Both women went into a new black Fiat 500.

Richard wanted to find out who these women are and looked at the back license plate. What he saw blew him away: the license plate appeared blurry. Richard could not make out the letters or numbers. Two days later, Richard looked into a mirror and could have sworn he saw the two women in black standing behind him and a voice in his head saying, "we expect your confidentiality." Richard was shaken to the core.

June 20, 2016

Rich Reynolds and the MIB

Given the series of events that began on this particular day, and which came to a head on the night of the 22nd, it would have been highly appropriate if the three day period was dominated by black clouds, tumultuous storms, and the proverbial dark and stormy night(s). In fact, it was quite the opposite: blistering temperatures and a pummeling sun were the order of the day. None of that, however, stopped profound weirdness surfacing from its sinister lair. And, here's how it all began.

On the evening of the 20th, good friend, Rich Reynolds, of the *UFO Conjecture(s)* blog, wrote a tongue-firmly-in-cheek article titled "Nick Redfern's latest book: The Women in Black." In part, Rich said:

> Nick Redfern's new book, *The Women in Black: The Creepy Companions of the Mysterious M.I.B.* may be read as a kind of sequel to his popular Men in Black books *The Real Men in Black* [2011] and *Men in Black* [2015]. (One could provide a psychiatric evaluation of Nick's obsession with this theme but that for another time.)
>
> As usual, Nick, in 292 pages, offers readers insights and hitherto unknown stories of odd appearances by women, often garbed in black clothing or accoutrements; appearances that scare the bejesus out of UFO witnesses, paranormal buffs, and normal people too.

June 21, 2016

"Some weirdo walking down the paved road in a black trench coat!!!!"

On the night following the one on which Rich Reynolds jokingly suggested I might need a psychiatric evaluation (at least, I *hope* he was joking!), I was scheduled to be interviewed by Denise Garcia, host of the *Phenomena Encountered* show. Denise wanted us to focus on the Men in Black and the Women in Black, which was fine by me. I should have guessed. Even before the show got started we had technical issues, with both Skype and Google Hangouts. Denise was forced to call me on my cell. She couldn't even connect to my landline (which, for those who may be interested in such things, ends in 666...). Within minutes people were messaging the show, and texting me, to say that the conversation was constantly dropping and there was endless interference on the line.

All of this came *immediately* after we had actually discussed how earlier shows on the MIB had been affected in such a fashion, such as *Coast to Coast* and Whitley Strieber's *Dreamland.* Finally, after about ten minutes, things settled down and a good, solid debate on the MIB and WIB followed.

Roughly an hour-and-a-half after I finished the interference-filled interview with Denise, cryptozoologist Loren Coleman and paranormal writer Brad Steiger both emailed me with the details of a breaking story coming out of Iowa, which just happens to be Brad's homestate. It was an article written by David Nelson, of KWQC news. Its title: "Mysterious 'men in black' sightings reported along Muscatine Co. roadways." And, in part, it stated:

> "R. J. Strong of Port Louisa, Iowa was spooked by what he saw a week ago Monday, June 13, at 2 a.m. Strong posted on Facebook afterward that he was traveling near Ogilvie and F avenues in Muscatine County when he saw 'some weirdo walking down the paved road in a black trench coat!!!!'"
>
> "Several others on social media have reported similar sightings in recent weeks. According to the reports, the individuals have been seen standing beside roadways and sometimes stepping into roadways as motorists are passing at locations across Muscatine County, in southeastern Iowa."

After that, things soon turned ominous. *Extremely* ominous.

13

"TWO MEN IN BLACK WEARING THE CLASSIC FEDORAS"

June 22, 2016

"Harrassed, scared, terrorized"

Rich Reynolds posted yet another article on my *Women in Black* book. Its title, once again with tongue-in-cheek, was "Are Nick Redfern's WIB witnesses nuts?"

Rich wrote:

"As some of you know, from reading Nick Redfern's latest book, *Women in Black,* an inconsiderable number of persons have claimed they were harassed, scared, terrorized by women in black, much as many UFO witnesses have been, allegedly, troubled by men in black. But are the encounters, noted by Nick in his 294-page book, real or hallucinatory? While persons seeing odd things – often having what is ascribed as a hallucination – they actually 'see' what they think they are seeing. But is there an actual, tangible 'thing' before their eyes?"

Rich's last question was an important one, as more than a few witnesses to the MIB have told me that they felt their experiences with both the Men in Black and the Women in Black occurred in a dream-like state – even something akin to the scenario presented in the 1999 movie, *The Matrix,* in which reality is far from what it appears to be.

July 23, 2016

Buffy Clary gets struck by lightning...yet again

The day was notable for two things; neither were in any shape or form positive in nature. With hindsight, I guess I could say that the whole weekend was weird. On Friday, July 22, good friend Erica Lukes invited me on her show to talk about my *Women in Black* book. The interview was postponed from the previous week, due to a huge storm that blacked out much of Salt Lake City, which is where Erica lives.

I didn't know it at the time, but this weekend was about to be dominated by thunder and lightning, too, and in a very sinister and almost unbelievable fashion. Friday was good: Erica and I had a fine time discussing the Women in Black, as well as my exploits with the *other* WIB – my various Goth girlfriends that have entered, left, and sometimes reentered, my life over the years. The atmosphere some forty-eight hours later or thereabouts, however, was very different.

It all began when Mark Henry asked me to come on his show – also to talk about the Women in Black and their male counterparts. All was going well until I brought up the matter of how both the MIB and the WIB were able to affect telephones. As has happened on so many other occasions, within about two minutes, the line went dead. Zero. Nothing. Nada. That is, until around five or six minutes later, when Mark was finally able to reestablish the connection. As was the case with so many hosts, Mark had never before experienced such a level of weirdness and bizarre technical issues as he had when I brought up the matter of phone interference.

Roughly twenty minutes later, the show was over. The high-strangeness, however, was not. Only a couple of minutes after Mark's show finished, Buffy Clary messaged me on my cell: She had just been struck by lightning. *Again. For the second time.* She was sitting in her yard when an almighty lightning bolt hit the big tree that dominates the yard, violently splitting it almost in two. As for Buffy herself, given that she was only around fifteen feet away at the time, she was shocked, felt tingly, weird, and very unwell. I suggested that she get someone to take her to the local emergency room – as in right now. She made it there, on her own, but couldn't get seen to for hours – despite shakily explaining to the staff what had just happened – and so she decided to head back home. It was two whole days before she felt herself again. Who can blame her for moving house not long afterwards? Not me, that's for sure.

July 19, 2016

"The MIB became a whirling blur of time and space"

Robert A. Goerman is an authority on the MIB. One of his reports appears in my *Men in Black* book of 2015. On this day, however, he sent me the following:

> Before you continue reading, let me preface this post by confessing that the MiB "visit" was merely a dream shard. I would never even publish my experience if not for its curious physical aftermath. It is also necessary to point out that I have resided at my present residence for much of my life. This was my childhood home. It was inherited upon the passing of my parents. The bedroom that I now occupy as an adult was at one time shared for many years with my younger brother. There was barely enough room for two

twin beds and a tiny writing nook where I documented my early research on a vintage Smith-Corona typewriter.

The dream that I had shortly after midnight on Monday, July 18, 2016 found me seated at my writing nook as a young man (1969 or 1970) at night. I was typing a report about the Men in Black and uttered to myself that these scenarios echoed the MIB cases reported by Brad Steiger. My desk lamp was the only light source in the room. Movement off to my left caught my attention and I turned my head. Two Men in Black, wearing the classic fedoras, stood not six feet away, between the twin beds, blocking any hope at escape. The closest one was about my height. His companion was a head taller. Both were no more than silhouettes. There were no details. Escape was the last thing on my mind.

I remember saying, "I have been looking forward to this," jumping up, and grabbing the shorter MIB by his arms between the elbows and shoulders. (Anecdotal information suggests that the MIB hate being touched by humans. I knew this.) BIG MISTAKE. It was like grabbing a high voltage tornado! In fact, the MIB became a whirling blur of time and space. My left arm was being ripped from its socket. I screamed that high-pitched scream that only excruciating pain can create: Pain beyond pain. Folks that have experienced certain traumatic accidents know exactly what I mean.

Hearing my cries for help, my wife awakened me. I snapped fully awake to the horror that the pain in my left arm and shoulder was quite real. (My firsthand relationship with pain comes from the agony of a herniated disc that required spinal surgery and too many kidney stones.) I understand pain. This was real pain. Maybe not quite as extreme as those two previous examples but still, as real as it gets. It took a healthy dose of Ibuprofen and the gentle working of the tender arm and shoulder to make things tolerable. Traces of soreness lasted most of the day.

It is what it is.

14

"HIS UPPER BODY AND HEAD WERE SILHOUETTED"

August 4, 2016

WIB, MIB and Werewolves

It was around 7:30 p.m. that Bigfoot-seeker Lyle Blackburn, cryptozoologist Ken Gerhard, psychic and Ouija-board expert Jen Devillier, and I headed off to the Ohio town of Defiance. A long drive was ahead of us. Around seventeen hours to be precise. But, we were a team on a mission. Earlier in the year, Ken had told me of his plans to hold a "Dogman Symposium," which was where we were now heading. For those who may be wondering, the term "Dogman" is, in essence, modern day wording for the world's most legendary shape-shifter, the werewolf.

When I bring up the matter of Dogmen, it very often results in the rolling of eyes and hoots of derision. The fact is, however, that there are hundreds of highly credible reports – from all across the United States, but particularly in Wisconsin, Michigan, and Ohio – on record of encounters with what can only be described as hair-covered, upright wolves. For the most part, these creepy critters provoke full-blown terror in the witnesses, as well as a sense that these entities are definitively evil. Linda Godfrey, undeniably the leading expert on the subject, has now written *six* books on the subject – such is the incredible body of data available.

UFO conferences, Bigfoot symposiums, and ghost-themed seminars are everywhere. Not so for the Dogman, however. That is, until Ken decided to rectify the situation by putting on an event specifically devoted to discussions of the dog-headed man-beast. It was a great idea, I thought. And when Ken invited me to speak at the gig I certainly wasn't going to say "no." Ken had a few concerns about putting on the event, but he shouldn't have. Hundreds of people booked tickets, all keen to learn the latest on – and the history of – this strange creature. There was a good line-up of speakers, including Stan Gordon, Linda Godfrey, John Tenney, and Black-Eyed Children authority, David Weatherly. But, I'm getting slightly ahead of myself.

As we headed out on the night of August 4 – Defiance-bound – I sensed this was going to be a weekend filled with intrigue. My instincts were not wrong. And there was a very good reason why Ken chose Defiance for the site of the first Dogman Symposium, as the following, brief diversion will make clear.

Midway through 1972, Defiance became what can only be termed "Werewolf Central." Over the course of a hysteria-filled two months – July to August – sightings of a rampaging, hair-covered man-beast, with a pronounced muzzle and dressed in rags, were made. The local media quickly picked up on the sinister saga, as did the town's police, who even opened an official file on Defiance's very own equivalent of Nessie, the Chupacabra and Bigfoot. To say that Defiance was gripped by terror would not be an understatement. Many of the sightings of the creature were made around a series of old railroad tracks, and usually late at night. A couple of guys working on the tracks – Ted Davis and Tom Jones – had an encounter of the very close kind. A close call, one might say. Davis told the local newspaper, *The Blade*: "I was connecting an air hose between two cars and was looking down. I saw these huge hairy feet, then I looked up and he was standing there with that big stick over his shoulder. When I started to say something, he took off for the woods."

For weeks, people were on edge. The Defiance Dogman was major news. And then, like a definitive specter of the night, it was no more. The legend, however, never really died away. The monster may be long gone, but memories of those days and nights of the early seventies still persist among those who lived through that brief, turbulent time of terror.

August 5, 2016

Monsters and MIB

After seventeen hours of taking turns to drive, sleep, eat, drink and listen to good music – fast, heavy and guitar-driven, for the most part – we were all ready to catch some shut-eye. No such luck! But it was all good. We got everything ready for the gig, and then headed off to a local restaurant to meet old friends, to make new ones, and to eat hearty dinners.

One of the more curious aspects of the MIB / WIB enigma – and which was discussed over dinner the night before the gig - is that the black-garbed fiends often turn up in locales where strange beasts have been seen. That's right: Contrary to what many might assume, it's not just UFO-themed cases that Men in Black and Women in Black turn up in. The most obvious example being the West Virginia town of Point Pleasant. When the fiery-eyed Mothman first descended upon the town from the dark skies above in 1966, the WIB and the MIB were never far behind. Similarly, in my 2015 book, *Men in Black*, I told the story of Colin Perks, a man who had a terrifying encounter with a WIB and a gargoyle-like beast back in 2000, in the ancient English town of Glastonbury. And, Stan Gordon has investigated cases where even Bigfoot witnesses have been plagued by the MIB, particularly so in late 1973.

It turns out that in 1967 – when Mothman mania was at its height in Point Pleasant – matters relative to the MIB were going down in none other than Defiance, too; very mysterious matters. It revolved around a man named Robert Easley – who was a UFO researcher and a good friend of Timothy Green Beckley, a long-time UFO sleuth and someone who, in the 1960s, managed to capture an MIB on film, in Jersey City.

For Easley, a resident of Defiance, it all began in the early hours of July 11, 1967. He was awoken by the sound of his phone ringing. Easley quickly answered it – perhaps fearing it was someone with bad news. It was actually a woman who wanted to report a sighting of two UFOs over the town – a sighting that was still going on. Easley jumped into his car and sped off into the night. It turns out that Easley was shadowed all the way by a black Cadillac, very often the preferred mode of transport for the MIB. The car had no license plates, which is typical too. As for the driver, who kept very close to Easley's vehicle, his upper body and head were silhouetted and dark. That was not good news. The MIB soon vanished, however, taking another road and leaving Easley worried and slightly paranoid. He would soon be back, though. On the night of July 15, the MIB returned, yet again in his black Cadillac, following Easley for a number of nail-biting miles.

Tim Beckley tells the story of what happened next: "When [Easley] pulled into his driveway, the unknown car sped off. Later that evening as he sat talking with his girlfriend on the front porch, the car came down the road and stopped right in front of the house, as soon as the topic of UFOs entered their conversation. Easley could feel the man looking at them. When they got off the subject the car left, but when they got back on it about an hour later, the same car came back again. It was as if the driver could hear what they were saying or read their minds."

All told, between the 11th and the 17th, Easley received twelve weird phone calls – all of which were of exactly the same kind. When Easley answered the phone, all he could hear were strange bleeping noises. It was the end of a very weird – and mercifully brief – period.

The matter of mysterious black cars, however, was far from over. While I was in Defiance I got fascinating reports from Linda Godfrey. It wasn't just fascinating; it was downright bizarre. *And it involved a Woman in Black and werewolves.*

August 6, 2016

A tale of a strange woman and a TV shoot

It was during a break at the Dogman Symposium on August 6 that Linda briefly mentioned her odd story; a story that she went on to expand as follows:

> It was around 2003, a sunny and warm day, and I happened to be out on Hospital Road, which is just off Bray Road, and where a lot of sightings of Dogmen have taken place – on that particular juncture of the road. There were a couple of cameramen and two colleagues of mine that were with me, and we were all being filmed for a TV show.
>
> It was the turn of one of my colleagues to be filmed. The other one and I were just standing at the side of the road – waiting and seeing what was going on – when this very large, black sedan pulled up. And as the window came

> down, there was an older woman in the car. Nobody that I knew. We looked at her as the window came down and she said, in an accent that sounded kind of like a Russian accent, but could have been eastern European; I'm not sure, "Do you need any help?" There was something about her mannerisms: it wasn't as if she was tentative or worried about anything. She just stated this in a really almost commanding way. And my friend and I looked at each other and we just said, "Thank you, no; we're just filming." And it was like she didn't really understand what we she said, because she said again, "Do you need any help? Can I help you?" We said, "No, thanks," again. She still didn't drive away; she looked over at the cameraman who was filming the activity and she seemed a little reluctant to leave. But, there wasn't any great reason she could think of to justify staying. So, she finally turned and stepped on the gas kind of slowly. And that was it.
>
> But, it just struck me as so strange at the time. She was dressed in black, but it wasn't a formal business suit or anything like that. The car was black and the interior was kind of dark. I would estimate she was probably in her sixties and she had a deep voice. Her hair was grayish and pulled back. There was a bit of exoticness to her. It was such an odd thing; it was out of the ordinary and stuck in my mind. She didn't fit in with the local populace, put it that way.

I couldn't help but think that the presence of the black Cadillac-driving MIB at Defiance, Ohio in 1967, followed five years later by a werewolf outbreak in town, and Linda's oddly synchronistic encounter with a WIB at Dogman-infested Bray Road, were all somehow connected. How? That question still bothers me.

August 6, 2016

A small town environment and a monster on the loose

The Dogman Symposium, dinner, drinks and good conversation were soon all behind us; the only thing that was left to do was to make that seventeen-hour-long drive back to Arlington. Poor Ken and Jen had another four hours to add on to that, for their even longer trek back to San Antonio. Very wisely, Lyle chose to fly back! But, before we all headed home, there was one thing left to do. The four of us – me, Ken, Jen and Lyle – drove out to the site of the very first encounters with the Defiance werewolf of 1972. Also along, in their own vehicles, were friends Denise Rector, John Tenney, and Amy Perry Lane.

As a fiery sun beat down upon us, we scoured the area and walked along the old railroad tracks. The area had changed very little since the 1970s – old photos of the area made that very clear. We mused upon what it must have been like – more than forty years earlier – to have encountered the Dogman of Defiance, late at night, on that dark stretch of track. It was, surely, chilling. A small-town environment and a monster on the loose: No wonder the legend had endured. And no wonder when, even in 2016, there was something oddly enticing about the prowling beast of the old railroad.

Before we left, me, Ken and Lyle decided it would be a very good idea to take a memento with us. We all chose the same things; several rusted, old railroad nails. When I got back home, I put the nail on one of my many bookshelves. And then I promptly forgot about it. Quite out of the blue, though, I later asked Kimberly Rackley – who is a skilled psychic intuitive – to see if she could pick up anything from handling the old nail.

I drove over to her apartment home in East Dallas on a Friday night – and I'm very pleased that I did. I deliberately didn't give Kim any clues as to what the artifact was, or where it came from, I just gave her the nail, sat back and watched, and let her do her thing. What she came up with was remarkable.

She said, "I see a shack with a painted roof. And, there's this big mark across the back of it, a long scratch." She then asked me, "Was this found at a crossing or at a railroad crossing?" I told Kim that, yes, it was found on a stretch of railroad track that crossed with a local road. In response, Kim said that she perceived some kind of supernatural phenomenon in the area. Notably, she added that it was as if the presence had been summoned – invoked or conjured up.

And she echoed, "I keep seeing that scratch and an image of claws on the building."

As I told Kim when she was finished, "That's really interesting, because where the nail was found, there actually is an old building next to it. And that you saw claws is interesting, because back in 1972, the railroad was the sight of a series of werewolf encounters in the area. And, the creature was reported scratching on the doors of local homes in the middle of the night."

I was absolutely sure that Kim had indeed made a connection and had picked up something deeply significant concerning the weirdness in Defiance. It was an amazing revelation.

15

"YOU COULD ALMOST TASTE THE MENACE"

September 2, 2016

Skinny and sinister

The Slenderman is a fictional character created in June 2009 by Eric Knudsen (using the alias of "Victor Surge," at the forum section of the *Something Awful* website), who took his inspiration from the world of horror fiction. The Slenderman is a creepy creature indeed; tall, thin, with long arms, a blank (faceless, even) expression, and wearing a dark suit, it sounds almost like a nightmarish version of the Men in Black of Ufology. While there is no doubt that Knudsen was the creator of what quickly became a definitive meme, people have since claimed to have seen the Slenderman in the *real* world.

So the theory goes, it's a case of believing in the existence of the Slenderman and, as a result, causing it to actually exist – which is very much akin to the phenomenon of the Tulpa or thought-form. An entity is envisaged in the mind, and to the point where the imagery becomes so powerful and intense that it causes that same, mind-based imagery to emerge into the real world, with some degree of independent existence and self-awareness. Such a scenario may well explain why people are now seeing something that began as a piece of online fiction.

One such witness to the Slenderman is Martine, who

encountered the Slenderman in her home in North Carolina on this date of September 2, 2016, and with whom I spoke just a few days after returning from a couple weeks of vacation back in England. It was late at night when Martine was disturbed in her sleep by what she described as a tall, thin, pale-faced man dressed in a black suit and who ominously whispered, "I can see you all the time," before vanishing into the darkness. Not the kind of encounter any of us should be hoping for.

September 16, 2016

Heading to Mothman territory

I have to say that September 16 provoked a distinct sense of déjà vu: me, Ken and Jen were off on another mammoth jaunt. This time it was to Point Pleasant, West Virginia – for the annual Mothman Festival. And, just like the Dogman Symposium of August 5, the journey was around seventeen hours long. But, we are all fans of road-trips, so it was a fun trek across the country. Plus, this was going to be a notable event as it represented the fiftieth anniversary of the beginning of the Mothman affair. We rolled into town – slightly blearily-eyed and frazzled from a lack of sleep – around 5:00 p.m., checked into our motel, and then, as darkness fell, headed out to a local pizza eatery. There were about twenty of us there, including John and Tim Frick, who I first met at the September 2014 Mothman gig. Not only do the Frick brothers know just about all of the Mothman intricacies inside out, but they also know the town itself very well, and particularly so the many and varied landmarks which are linked to the story of the famous, winged monster. To the extent that later that night, around 8:00 p.m., they took us – along with other friends, including Denise Rector and Audrey Hamilton – to the old, so-called "TNT area," which played such an integral role in the Mothman saga and that of the MIB, too. As we left the pizza place, a full-blown convoy of vehicles followed Tim and John as we traveled along small, winding, tree-shrouded, roads to the scene of the old and eerie action. No surprise, there was an atmosphere of excitement and anticipation in the increasingly chilly air.

For those who may not know, the TNT area's official title is the McClintic Wildlife Management Area. It's situated around five miles north of the town of Point Pleasant and runs to more than 3,500 acres. At the height of the Second World War, a TNT processing plant was established in the area, with the volatile chemicals used to create it stored in a series of concrete, igloo-like buildings. It was the work of around 3,500 U.S. Army personnel and, at the time, was known as the West Virginia Ordnance Works. Today, the plant is no more. The only things left now are the crumbling foundations and a couple of sturdy, metal, perimeter-gates and a rusted metal fence. During daylight, it's an inviting and picturesque area, filled with densely packed trees, a plethora of wildlife – such as deer, raccoons and beavers – and numerous ponds, pools and small lakes. After dark, however, things are very different. The atmosphere of menace, which was so present back in the sixties, is *still* there – utterly refusing to relinquish its icy grip on the people of Point Pleasant.

Having checked out what was left of the old plant, we all followed John and Tim to a specific stretch of heavily wooded ground, parked our vehicles, and were given an excellent and atmospheric tour of the igloos and their surroundings. I have to say the whole thing reminded me of something straight out of *The Walking Dead*: a ruined, overgrown environment, a once bustling area now utterly dead and abandoned, and an almost apocalyptic air that one could practically cut with a knife. The military was nowhere in sight, and the igloos were decaying, covered in foliage, and splattered with graffiti, both old and new.

Denise and I broke off from the main group and checked out some of the igloos, which was a profoundly memorable experience: the size and shape of the igloos causes a person's voice to echo loudly and very oddly within their dark confines. Plus, we felt a deep sense of malignancy in the old buildings – a sense that was as immediate as it was long-lasting. You could almost taste the menace, if such a thing were possible.

Notably, there was some evidence that supernatural rites and rituals had been undertaken in some of the igloos, which I found most intriguing. It was near these very same igloos that so many of the Mothman sightings occurred in 1966 and 1967 – involving, it should be noted, witnesses who soon found themselves in the cold clutches of the Men in Black. Denise and I walked around, in near-darkness, for a couple of hours, with little more than the bright Moon for illumination, taking in the atmosphere and imagining what it must have been like here fifty years earlier. A few scurrying animals and the cries of a handful of geese flying overhead were pretty much the only things that convinced us we hadn't entered some strange portal – a doorway to an unsettling, dead world. After a while, we caught up with the rest of the gang. We hit the darkness-filled roads and headed back to our motels. It had been a cool night of high-strangeness.

September 17, 2016

Tales of the MIB

Men in Black are roaming the streets of Point Pleasant. Downtown is cordoned off. Sightings of a terrifying, glowing-eyed, winged creature abound. Nope, we're not talking about one of those awful, cheap movies that the SyFy Channel insists on regularly bombarding us with. This is the Mothman Festival. At the time of this writing, it's in its sixteenth year, and is the highlight of the calendar for the people of the town and its surroundings. Back in the 1960s, sightings of the Mothman caught the attention of John Keel. A skilled writer, Keel, in 1975, penned a book on the mystery titled *The Mothman Prophecies*. In 2002, it was turned into a big-bucks movie of the same name starring Richard Gere and Laura Linney.

The success of the movie prompted the city to hold a yearly celebration in honor of the freaky, flying whatsit. And, yes, the streets really *are* closed down, people *do* dress up as the dark-suited characters made famous by Will Smith and Tommy Lee Jones, and you're all but guaranteed to see more than a few people disguised as Mothman. A huge, aluminum statue of the winged thing, made by sculptor Bob Roach, dominates downtown and is a monster-mecca for attendees. Local rock bands offer a fine barrage of tunes. There's a pageant for those girls who might want to be crowned "Miss Mothman." And, of course, there are the obligatory Mothman fridge-magnets, t-shirts, and other assorted mothy memorabilia.

Ken, Jen and I had an early start the next morning – specifically to set up our book-tables at the downtown, open-air festival. I looked out of the hotel-room window: the sky was grey and dark and the rain was hammering down. We weren't daunted, however:

we were all set up by 8:00 a.m. To say that it was a successful day would be an understatement of epic proportions. Slightly more than *ten thousand people* descended on the town and the festival. Yes, *ten thousand.* One of them was Dennis Carroll, who had shared with me his MIB-themed encounter earlier in the year. It was a strange affair that was still on Dennis' mind, months later. No surprises there: it's hard for anyone who has seen the MIB to ever completely forget about them.

One of the people who came to chat with me, as I sat at my vending table, was Irene, a woman who related a very bizarre story; one with clear MIB / WIB overtones attached to it. As Irene revealed, she is now retired and had a very strange experience while living in Pasadena, California in the late 1970s – either 1978 or 1979. In the early hours of the particular morning in question, she inexplicably woke up and felt compelled to go to the bedroom window. On doing so, she was shocked to see a car-sized, oval-shaped UFO – of a bright blue color – that hung in the air for a few seconds, before vanishing into the dark sky above. She raced back to the bed, terrified by what she had just seen. On the following evening, however, something even stranger occurred.

There was a knock at the door. It was a very pale-faced woman in black who claimed to work for a company that offered aerial photographs of peoples' homes – taken by an expert photographer, then framed, and that would look very nice on the living-room wall. There was, however, something that made Irene think things were just not quite right, not at all. Her instinct proved to be right on the money. When the woman, somewhat oddly, offered to show Irene one of the cameras that the company used, she agreed to see it – although why she would need to see it at all makes very little sense. If any sense at all. The woman proceeded down the drive and got in the backseat of the black car she had arrived in. In a few seconds, she was out of the car. Irene could see the camera, but was very surprised by what happened next. The woman quickly took a photograph of the front of Irene's house, jumped back into the car, and in seconds was out of sight.

Irene quickly became convinced that the odd encounter was somehow linked to the events of the previous night. She was likely not wrong. John Keel called these odd characters "Phantom Photographers." In his classic book, *The Mothman Prophecies*, Keel said of one such case: "On a rainy night in April a man from Ohio had been driving along Route 2 near the Chief Cornstalk Hunting Grounds when a large black form rose from the woods and flew over his car. 'It was at least ten feet wide,' he claimed. 'I stepped on the gas and it kept right up with me. We were doing over seventy. It scared the hell out of me. Then I saw it move ahead of me and turn toward the river.'"

Keel continued: "Months later, late in October, he returned home from work and found a prowler in his apartment. 'When I opened the door I saw this man standing in my living room,' he reported. 'I think he was dressed all in black. I couldn't see his face, but he was about five feet nine. I started to fumble for the light switch when he took my picture. There was a big flash of light, so bright I couldn't see a thing. While I was rubbing my eyes the burglar darted past me and went out the open door. I guess I arrived just in time because nothing was missing.'"

When I told Irene of Keel's work on the issue of the Phantom Photographers, it at least helped her to realize that she was not alone and was not crazy. She went away relieved, but not entirely comfortable, which hardly surprised me. Irene thanked me, waved goodbye, and was quickly swallowed up by the huge throng.

The rest of the Mothman Festival went phenomenally well – friendships were forged, tales were told, and the legend of the Mothman grew evermore.

16

"SHE HAD TOTAL BLACK EYES"

September 21

Heading out to Sin City

Tracie Austin is a long-time friend of mine who I first met back in 1995. At the time, Tracie and I only lived about a 45-minute-drive from each other in central rural England. As a result, we spent a lot of time traveling to UFO conferences in the U.K. and having a good time, very often with a mutual friend, Irene Bott, who used to run the Staffordshire UFO Group. Tracie moved to the United States a couple of years before me and now lives in Las Vegas, Nevada. She has done very well for herself and has her own TV show, which is broadcasted out of the heart of the city.

In the summer of 2016, Tracie invited me and David Weatherly to be on her show. The plan was to have Tracie first interview us separately: I would be interviewed on the MIB and David on the Black-Eyed Children. And, then, there would be a roundtable discussion on both phenomena, noting the undeniable links between the two. A date was soon fixed: September 22. Both David and I flew in on the same morning and spent the afternoon hanging out at our hotel.

The high-strangeness wasn't just limited to the show, however. Tracie herself had a trio of stories of her own to relate to me. They were, of course, as bizarre as they were disturbing. She told me:

> With me doing the show on the Black-Eyed Kids, I was reading David's book every day before we did the show. So, it was on my mind. And it only happened the one time, but I had this dream that we had a friend over and there was a knock at the door, and the friend that came over opened the door. And I saw a black-eyed kid standing at the door; one single boy. And the friend asked him to come in, and I said, "No, no, no; you can't let him in!" But, he wandered in, slightly. It was a kid with a hoodie and jeans; the typical kid. And I pushed him out the door and said again, "No, you can't let him in!" It shook me a bit when I woke up.

The skeptic might be inclined to say that because Tracie was focusing on the BEC at the time, and diligently preparing for the show, that all of the data she had digested had now spilled over into her dream-state and her subconscious. I knew that such a thing was not at all implausible. I also knew, however, that the MIB had the uncanny ability to, essentially, invade our dreams and turn them into absolute nightmares – as we have seen in previous chapters. Maybe, the Black-Eyed Children could do likewise, too, I considered. That was far from being the end of the story, however.

Tracie continued with what she had found out:

> I have a friend named Dana; she works in timeshare, at a resort out here in Las Vegas, on the far end of the strip. She was working one day – this was in April 2016 – when she said this lady was sent over to her, with two kids: a young boy and a girl. The girl was about six or seven and the boy was about four or five. The woman was wearing dark brown pants and had long brown hair. And the reason

why the woman was sent over was because Dana gave the gifting in the timeshare company. When you listen to a presentation, you're given a gift. But, the woman said she didn't want a gift and Dana said, "When she looked at me and said that, I saw that she had total black eyes." There were no whites whatsoever. And the kids were exactly the same. But, in every other sense they were completely normal; they weren't wearing the hoodies.

Dana commented, "I wish I had said something, like how do you see with those eyes?" When she said that, I thought, you can't really ask them something like that. Maybe you could say, "Wow, your eyes are beautiful." But, maybe that's bringing attention to something they don't want you to know about. But, they could wear sunglasses all the time and no one would know. So, maybe they are testing our reactions, they want their eyes to be seen. The kids were quiet; they didn't say anything to her. Dana asked the woman how long she was going to be staying at the resort. And she said just one night and then they left.

Dana said to me, "I couldn't believe it, Trace; I saw after that that you were going to be doing a show on the Black-Eyed Kids." She reached out to me on Facebook and said, "I had an encounter with a Black-Eyed Woman and Black-Eyed Kids." Then, a couple of months after encountering this woman, Dana asked me if there was some connection between seeing them and people dying, when they have an encounter with the Black-Eyed-People. Right around a month or two after her encounter, her parents – who were both in their seventies – died. They hadn't been sick, weren't ill. Her father had developed kidney failure, out of the blue, and then her mother developed pneumonia. She said to me, "Tracie, this is freaking me out, as I hear

> [the Black-Eyed Children] can be omens of death." I said I didn't know, as I didn't want to scare her.

And there was one more story from Tracie:

> I also have a friend who had an experience with a Black-Eyed Man; this was in a Wal-Mart in Michigan where she saw him. I was living in California at the time and this would have been about 2004, Christmas time. They were at the checkout and he insisted that he gave a gift to her son. It was a Spiderman clock. She said that where he stood in line, there was someone in between her and this man. But, when he gave the gift, it was like his hand reached through the person between them; just passed right through the flesh of this person. And what freaked her out was that he had the pure black eyes. She didn't want him to accept the clock, with him being a stranger. But, he insisted.

So far as Tracie knows, the strange gifting – from an even stranger man - did not provoke any kind of supernatural backlash. *Yet.*

My apartment: the scene of so much mayhem and weirdness. (Nick Redfern)

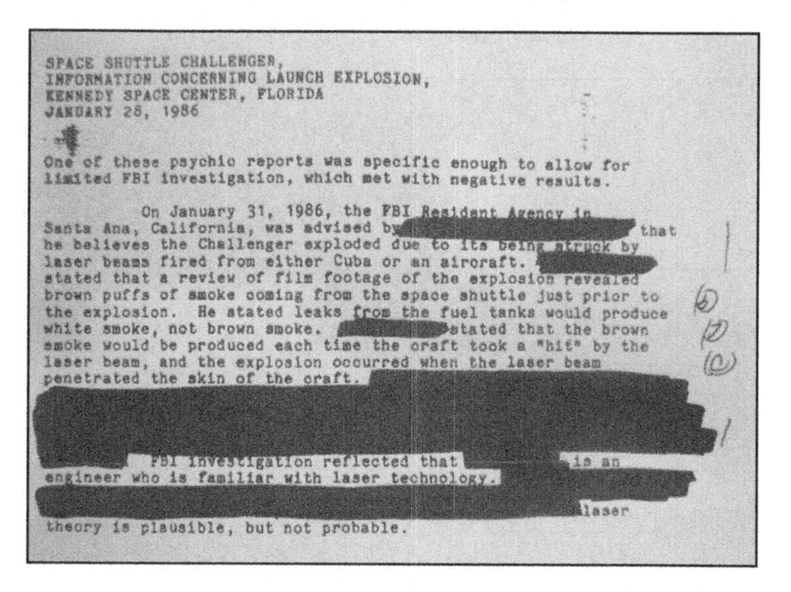

SPACE SHUTTLE CHALLENGER,
INFORMATION CONCERNING LAUNCH EXPLOSION,
KENNEDY SPACE CENTER, FLORIDA
JANUARY 28, 1986

One of these psychic reports was specific enough to allow for limited FBI investigation, which met with negative results.

On January 31, 1986, the FBI Resident Agency in Santa Ana, California, was advised by that he believes the Challenger exploded due to its being struck by laser beams fired from either Cuba or an aircraft. stated that a review of film footage of the explosion revealed brown puffs of smoke coming from the space shuttle just prior to the explosion. He stated leaks from the fuel tanks would produce white smoke, not brown smoke. stated that the brown smoke would be produced each time the craft took a "hit" by the laser beam, and the explosion occurred when the laser beam penetrated the skin of the craft.

FBI investigation reflected that is an engineer who is familiar with laser technology. laser theory is plausible, but not probable.

An extract from the FBI's file on the 1986 *Challenger* Space Shuttle disaster. (FBI)

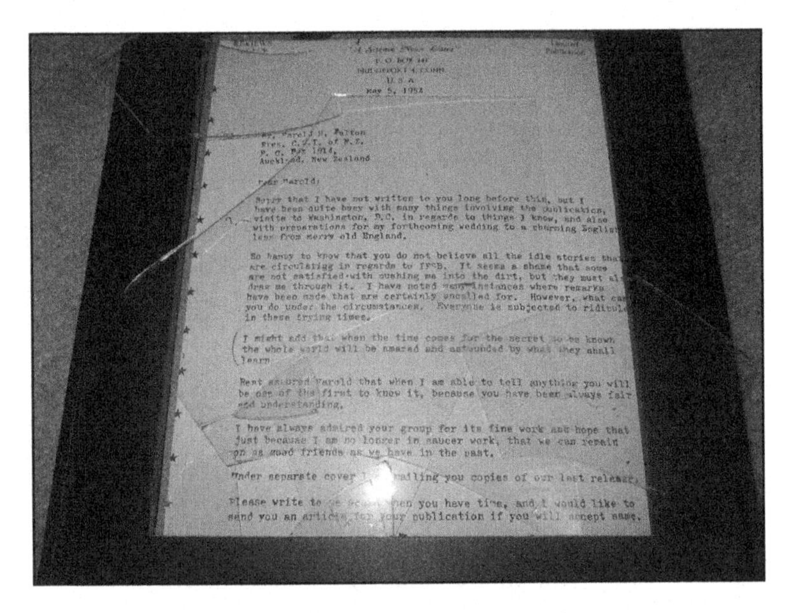

P.O. BOX [illegible]
[illegible] CONN.
U.S.A.
May 5, 1954

Mr. Harold H. Fulton
Pres. C.S.I. of N.Z.
P. O. Box 1914,
Auckland, New Zealand

Dear Harold:

Sorry that I have not written to you long before this, but I have been quite busy with many things involving the publication, visits to Washington, D.C. in regards to things I know, and also with preparations for my forthcoming wedding to a charming Englis[illegible] lass from merry old England.

So happy to know that you do not believe all the idle stories tha[illegible] are circulating in regards to IFSB. It seems a shame that some are not satisfied with pushing me into the dirt, but they must al[illegible] drag me through it. I have noted many instances where remarks have been made that are certainly uncalled for. However, what ca[illegible] you do under the circumstances. Everyone is subjected to ridicul[illegible] in these trying times.

I might add that when the time comes for the secret to be known the whole world will be amazed and astounded by what they shall learn.

Rest assured Harold that when I am able to tell anything you will be one of the first to know it, because you have been always fair and understanding.

I have always admired your group for its fine work and hope that just because I am no longer in saucer work, that we can remain on as good friends as we have in the past.

Under separate cover [illegible] mailing you copies of our last releas[illegible]

Please write to [illegible] when you have time, and I would like to send you an article for your publication if you will accept same.

Shattered: a framed letter written by MIB chronicler Albert Bender. (Nick Redfern)

Welcome to Defiance: from MIB to Dogmen. (Nick Redfern)

Lyle Blackburn, John Tenney, me, and Ken Gerhard: hanging out in Defiance, August 2016. (Nick Redfern)

Beware of the Mothman. (Nick Redfern)

One of Point Pleasant's old and atmospheric "igloos." (Nick Redfern)

Buffy Clary: not a fan of electricity! (Nick Redfern)

These two individuals also went to the Town Tax Map Department and viewed diagrams of subject's property and adjacent property and asked several questions about ARMSTRONG in that department.

b6
b7C

According to [redacted] no one really questioned these two individuals, although they were suspicious and these individuals stated they had been in town just to take some photographs of the house as they were tourists.

[redacted] advised the only description he was able to get was they were both fairly young and male, and well-dressed. The Negro appeared to have a type of necklace, which had a quarter moon with a star on it. No one observed what type of automobile or any other details about these individuals.

[redacted] advised that on the following day, [redacted] came to the Town Hall and was quite concerned as to these two individuals and made a statement to one of the town employees, "I didn't think they would go this far".

7-0-115

1-Cincinnati

MJD:pl
(1)

161- 18507-

An FBI paper on Neil Armstrong's mysterious visitors. (FBI)

From *Rosemary's Baby* to a Man in Black. (Nick Redfern)

MICROWAVE
TECHNIQUES

NAVSHIPS 900,028

Prepared by

NDRC—Division 14
RADIATION LABORATORY
Massachusetts Institute of Technology

Published by

BUREAU OF SHIPS NAVY DEPARTMENT

The Philadelphia Experiment: a document left on the doorstep. (Nick Redfern)

Hanging out with Susan Sheppard at the Mothman Festival, 2016. (Nick Redfern)

Tracie Austin, who shared her Black-Eyed Children stories with me. (Tracie Austin)

Dallas, Texas' "M.I.B. Grave."
(Nick Redfern)

Denise Rector walks the old cemetery.
(Nick Redfern)

17

"ONE OF THE MIBS TURNED AND STARED INTENTLY AT HER"

October 28, 2016

Old reports of the MIB surface

I was spending the weekend at Kent Senter's *Night Siege* conference in Greensboro, North Carolina. It was a fun gig and one that was well-attended. My agent Lisa Hagan was there too, as was her mom, Sandra Martin, who is the brains behind Paraview Books. I hadn't seen either of them for a few years, so it was great to catch up again. Also speaking at the gig were Travis Walton and Thomas Reed. It was during a quiet few moments in the event, and before I was due to do a presentation on the MIB/WIB issue, that a man came up to me and gave me a manila envelope. He introduced himself as Wayne LaPorte.

I have to confess that I didn't recognize his name. As we got talking, however, it became clear to me that Wayne had been a significant figure in 1970s-1980-era U.S.-based Ufology, having written for a number of UFO-themed magazines and journals back then. I opened the envelope to find a couple of MIB-based cases that Wayne investigated in the 1970s. He very generously let me keep the reports; he also gave me permission to reproduce them. One of them was written to Wayne by a now-retired employee of the *Gaston Gazette* newspaper, which covers the town of Gaston, North Carolina. Dated July 23, 1977, it read:

As a part of my work as a reporter for the *Gaston Gazette*, I wrote a number of stories during the last three months of 1976 that dealt with UFO sightings in Gaston County. I think it is important for anyone who reads this recollection to know that I was interested in these stories mainly for their sensational news appeal. I have never attempted to make a study of UFOs or occupants and know virtually nothing about the subject.

During the Oct.-Dec. 1976 period in which I was writing stories about hundreds of UFO sightings in Gaston County, I received a telephone call from an unidentified woman. I believe the call was made sometime in November. I do remember it was on a Saturday afternoon between 4-6 p.m. The woman would not reveal her identity because, she said, too many people would think that she was "crazy."

I pushed for a story during the call that lasted about 45 minutes, but the woman – she sounded middle age – would not consent. In fact, it was her really strong desire to remain unidentified, even to me, which made me believe she was serious about what she was saying.

My recollections are vague, but I do remember her beginning her story by telling me about UFO visitations to her home. She would not tell me where she was located, other than in Gaston County. She said something about strange noises outside her bedroom window, rocks being thrown on the windows. The noises caused her dog to bark continuously.

During the several times that she believed UFOs to be outside her home, she and her mother and child would close themselves in a bedroom, the same bedroom I referred to earlier. I remember her telling me that she read the Bible when she heard the noises and that the reading seemed to drive the noises away.

At some time after she experienced the visitations, a stranger came to her home. She told me the name he used but I do not recall it. He told her he was from Chapel Hill and she said that he had some knowledge of the Chapel Hill area.

I believe she told me that the man had blond hair and wore a black turtleneck sweater or knit shirt. The turtleneck part I am sure of. She also described him as having a larger-than-normal head.

She and her husband talked to him in their home for a while then took the man to Shoney's for a meal. She told me that he held the menu upside down and did not know what to order. She ordered fried chicken for him. He told her, she said, that he did not know what fried chicken was. He ate it anyway she said.

That's about all I can remember her telling me about the visit.

But she said that later she tried to call him in Chapel Hill. There was no listing for the name he gave.

I remember that she told me the reasons she believed he was an extra-terrestrial being, but I do not remember what those reasons were. I can't be sure, but it seems that she said he left an address and that she wrote to him at the address and the letter was returned.

I begged her to tell me if she ever had another similar experience. She promised to call me. She has not and I do not think she ever will because she knows that I want her story and she is 100 per cent against even a no-name type story.

I have been a newspaper reporter since 1969 and I have never talked to an unidentified caller who, I sensed,

> was so careful to give no hints about her identity or her residence. And at the same time, I felt she was completely convinced about her experiences. In any event, I am convinced that she was not playing a prank to see if we would bite for a story.

As the dossier of material that Wayne provided to me makes clear, he made his own handwritten notes on the events as described above, and around the time he received the report. Wayne wrote:

> This incident was first related to me by [deleted] in November 1976 when I first met with her to discuss the reported UFO calls. [Deleted] said, "This is the woman you need to talk to..."
>
> The only other points that I recall [deleted] telling me that are not covered above are:
>
> 1. The first experience was when a UFO came and hovered near her home. It went away after a short time.
> 2. She told the neighbors and they laughed at her. As a result she did not telephone the police or any government official.
> 3. A few days later a UFO investigator (the MIB) showed up. This puzzled her as she was astonished that the government had found out about here UFO sighting. She concluded that her neighbors must have told someone who knew someone who put this investigator in contact with her.
> 4. I don't recall a time span but it seems like she had three visitations over a period of a few years. The MIB visit happened shortly after the first visit or encounter.

> The other important item not mentioned above is the purpose of the call. [Deleted] is a skeptical believer and slanted her stories in a humorous manner. This humorous manner upset this "mystery caller." The purpose of the call was to convey to [Deleted] that UFO's were serious business – especially to some people such as herself (the caller). Either this person is serious or well read on Ufology (especially Keel). However, the purpose and lack of desire for publicity makes me feel this incident is real.

I read the report while waiting to do my lecture and I realized that this particular saga was ripe with MIB overtones: the black clothing, the curious shape of the head of the mysterious visitor, his baffling ability know that the woman had had UFO experiences, and the MIB's inability to grasp the concept of food, have all been reported time and again. It was satisfyingly appropriate that in mere minutes I would be on-stage, talking about darkly-clothed, strange entities – one of who appeared to have invaded the space of a puzzled and disturbed family from Gaston County, North Carolina, around forty years earlier.

October 29, 2016

"The MIBs themselves were an energy entity"

As I took to the skies on Sunday morning, and with the *Night Siege* gig having been a resounding success, I sat back in my seat (at least, as far back as today's aircraft will allow you to stretch out) and read the second story from Wayne LaPorte:

> This account was first related to me by two friends of mine – Don and Sarah Morris [address deleted]. Both have read

> little on UFOs, and have read nothing on Men in Black (MIBs). On May 27, 1977, Don, Sarah and myself had supper out. I was discussing hostility sometimes expressed by UFOs. I mentioned that there are on the record cases of Men in Black or MIBs. They asked me what that meant. I said, 'The MIBs may be an enforcement arm of the UFOnauts. They often dress in black, and are very human-like.' Sarah said: "Really? We had a strange experience with some men dressed in black turtleneck sweaters at a Kent State Memorial Service in 1972. We photographed them, because of their dress. Our film was ruined. There's not a UFO involved here. However, it was very odd that our film was ruined.

I got myself a can of Heineken and a bag of chips from the flight-attendant and continued to read LaPorte's report. It was, perhaps, appropriate that as I got to the next part of the story, the heavens turned grey and we hit a spate of significant turbulence. *Nightmare at 20,000 Feet?* Nah, but amusing to think about it! LaPorte wrote:

> It was the first part of July, 1972. Don and Sarah attended a Kent State Memorial Service at Freedom Park in Charlotte, N.C. Both are amateur photographers. They had a Honeywell Pentax 35mm camera. The service was a non-violent, anti-war demonstration. There was a stage set up. Various people were giving speeches. There were some singers and group singing also. It was sunny and very hot.
>
> The FBI were there and very obvious. They were going through the crowd and photographing various people mainly for harassments. Don wasn't a protestor (active), and did not have a photographic file on the FBI. However,

he was upset that the FBI was bothering the people. So, he decided to harass the FBI by taking photos of the FBI photographers.

Don and Sarah first noticed the MIB when they first drove into Freedom Park (a recreational park a few miles from downtown Charlotte). They were in the auto in front of Sarah. One of the MIBs turned and stared intently at her. She noticed the turtleneck shirt and they were clean shaven. They had short hair. She said, 'They must be part of the local FBI showing up.' She thought they were FBI, because of the short haircut. The two MIBs and the FBI men were the only ones with short hair at the Memorial Service.

Don and Sarah later noticed the two MIBs. Don had been walking around photographing the FBI agents. He returned to Sarah who was with the audience. A simple stage had been erected. The audience was seated in front of it and spread out like a fan. Don and Sarah were near the back of the audience. There were no chairs. Everyone was seated on the grass.

The two MIBs stood behind the audience and about 50 feet from the audience. They wore black turtleneck shirts or sweaters (both can't recall whether they had on shirts or sweaters). Their clothes were identical – also very clean and neat. They never talked (even to each other). They never went to eat or to the restroom. They stayed in the same spot all day. The temperature was about 27 degrees centigrade. The MIBs weren't sweating. They just kept looking around, staring at everyone.

Don and Sarah can't recall any facial features. However, they never thought of them being non-human. They thought the MIB were "FBI supervisors." They took a few photos of them from the crowd. Then, Don went around and came up behind them. He got to five feet (1.5 meters) behind them. He focused the camera. He hoped they would sense he was behind them and turn around. He wanted to capture the surprise on their faces. However, he waited behind them for five minutes. They *never* turned around. Then, like he was "commanded," Don returned to Sarah in the audience. He doesn't recall any verbal or mental command. However, all of a sudden, he lost interest in trying to photograph the MIBs.

The next week, Don and Sarah developed the black-and-white film at their home photolab. The film was completely black from front to back. It was as if struck by some form of intense electromagnetic radiation. Their thoughts were that the FBI had a secret radiation weapon used on them to destroy their film. However, Don doesn't recall seeing the FBI or MIBs point any weapon at them. Also, Don doesn't recall experiencing any radiation sickness.

There is the possibility that Don had purchased a roll of film that had been accidentally exposed to radiation while being shipped to the retail store. To me, that is possible, but very unlikely. *If the film was good,* there might be several reasons for the blanks: 1. The FBI have a secret radiation weapon; 2. The MIBs used some type of weapon secretly on Don's camera; 3. The MIBs psychically were able to destroy the film; 4. The MIBs themselves were an energy entity or were radiating electromagnetic radiation; or 5 The MIBs had a protective energy field around them.

Harry Joyner owns a photolab (Joyner and Associates) in Charlotte. He offered to give us free photographic assistance on any case. He listened to the taped interview. Harry said the description of the ruined film indicates

> the following. Total exposure to light or intense heat or chemical fumes or exposure to gamma, alpha or beta radiation. The witnesses said they made no darkroom mistakes while developing the film.
>
> Harry also does some photographic work for the FBI. He told (at my request) on the FBI agents of the MIB incident. He asked the agent if the two MIBs could have been FBI agents. The agent replied, "No." The agent wasn't involved in the 1972 Freedom Park affair. However, he said that the FBI would never dress like that, and act in such a manner (not moving around, not talking, etc.).
>
> Although no UFO is involved here, the two Men in Black appear to be MIBs. Nolie Bell of the Tarheel UFO Study Group, Winston, Salem, N.C. knows of a similar case. Two MIBs were spotted at a Kent State Memorial service. That report said the MIBs had oriental features.
>
> Also, John Keel related to me that he knows of one case in which possible MIBs were photographed. A photograph was taken of a group of (outdoors) attending a meeting of a radical nature. There were two men in different parts of the crowd. They had on black suits with turtleneck shirts. They both looked like twins. They had lean, angular faces. The hair was cropped short. This was a group photo. It was not a close-up. Incidentally, this was in England.
>
> I feel this case is significant as it involves a close-up attempt to photograph a UFO.

When I read the latter account – of both the MIB and the FBI being present at the scene in question – it instantly made me think of something John Keel once wrote on the matter of *two different kinds* of Men in Black. Keel put down on paper, on March 15, 1968 the following: "I have very substantial reasons for believing that the current explosion of MIB activity is directly related to a forthcoming national crisis. The crisis has been developing over a long period of time. The numerous hoaxes, deceptions, and seeming meaningless incidents inherent in the UFO phenomenon have actually served to cover up the real situation."

Now, we get to the crux of the matter. Keel added: "*It appears as if two groups of MIB are involved* [italics mine]. One group is extremely dangerous. They have committed murder and arson on a broad scale, operating under the carefully constructed umbrella of ridicule and nonsense surrounding the UFO phenomenon.

"The second group is trying to focus our attention on the first group by imitating MIB activity through harmless interviews, presentation of obviously false credentials, etc. We must understand and appreciate the efforts of this second group, and learn to discriminate between the two."

I thought, Keel was *always* ahead of his time. In whatever afterlife domain he now dwells, he probably still is.

18

"THEY WERE PALE AND SICKLY"

November 2, 2016

Black helicopters that perhaps aren't

Mysterious Universe published a new article from me. Its title? "Black Helicopters: What in Hell?!" The article was on a very odd subject; that of the so-called "Black Helicopters" and "Phantom Helicopters" that are so often seen at the sites of notable UFO encounters and cattle-mutilation events. For most researchers, the helicopters are almost certainly the property of clandestine agencies and programs that use the craft to keep watch on the UFO phenomenon and its activities. There is, however, a stranger side to the mystery than that; a profoundly weird side.

I have more than twenty reports in my files from people who claim to have seen black helicopters – flying silently – that morphed into the shapes of classic 1950s-era flying saucers, into small balls of light, and into large, blinding balls of light. Such reports stretch credulity to the max. But, they are widespread. So what is going on? I don't have the answers to that question, but, here is one such case, from my files, which dates from February 1982.

The town of Dulce, New Mexico has become infamous in UFO circles, chiefly because it is said to be the home of a massive underground base – which is manned by hostile, dangerous aliens who engage in nightmarish genetic experiments on human abductees. So the sensational story continues, the U.S. Government knows all too well what goes on under Dulce, but lacks the sufficient power or ability to eject or destroy the alien hordes. That there has been an undeniable mass of so-called cattle-mutilations in the area is an issue that cannot be denied – indeed, the FBI has declassified a sizeable number of reports of such mutilations from the area.

This brings us to the story of Bruno. On the day in question, he saw a squadron of black helicopters hovering over Dulce's Archuleta Mesa – below which, the story goes, is the extraterrestrial installation. As Bruno watched, the helicopters inexplicably morphed into bright balls of light – leading him to conclude that the E.T.s of Dulce can camouflage their craft to resemble our aircraft, thus ensuring that, for the most part, they are never seen for what they really are.

Another answer, of a *very* strange sort, surfaced in 1994. Back in 2007, I interviewed Ray Boeche, a priest and a former Mutual UFO Network (MUFON) State-Director for Nebraska. In 1993, Ray met with a pair of U.S. Department of Defense physicists who were working on a classified program to contact what many would call aliens. The program, however, referred to them as Non-Human Entities, or NHEs. As the project continued, and as runs of bad luck, ill health and even death blighted the research, the members came to believe that the NHEs were not E.T.s, after all, but literal demons from an equally literal Hell.

In 1994, those same two DoD physicists stated: "Regarding the phantom helicopters, while many are direct NHE 'productions' (craft is not an appropriate term as they do not need to travel via a propulsion device) many are related to our program, especially regarding running checks and surveillance on mutilation sites and so-called abduction victims."

After the article was published, several people contacted me to say they too felt the black helicopters they encountered were unearthly and evil hologram-style creations. One of those was Mothman expert John Frick, who shared with me his thoughts on this issue. I didn't know it at the time, but John's account of a personal encounter with a black helicopter spilled over into other issues, too, including that of not just one Woman in Black, but of *two.*

I arranged to speak with John on the dark night of November 7, and which I did. Before I could do so, however, and little did I know it at the time, a very bizarre and surreal series of events was about to erupt across November 4 and 5.

November 4, 2016

Rosemary's Baby - "something" came calling

It was on this day that I penned an article titled "Rosemary's Baby: The Controversy Continues." The subject matter was a weird one – as most of them tend to be! Namely, the many and varied examples of how people had experienced disturbing supernatural phenomena after watching the 1968 movie, *Rosemary's Baby* – or after reading Ira Levin's 1967 novel on which the movie itself was based. Even the soundtrack to the movie plays a role in this sinister affair. As an aside, this is not unlike the situation which David Weatherly and I found ourselves in, when the very act of people reading our books seemed to provoke supernatural manifestations. Now, back to the matter under the microscope.

The two key stars in *Rosemary's Baby* are actors Mia Farrow and John Cassavetes. They play a married couple, Guy and Rosemary. When they move into a new apartment, all goes well. Or, rather, it all goes well for a while. It's not long before things start to go menacingly wrong. Very wrong. We're talking about neighbors who are not what they appear to be, demonic activity, a secret group of Satanists, and – at the heart of it all – terrified Rosemary, whose pregnancy very quickly becomes integral to the plot. And, it scarcely needs saying, so does the baby.

One of those who had a very bizarre *Rosemary's Baby*-connected tale to tell was an actor and voice-over artist, Peter Beckman. As Peter told me, the experience occurred in either 1969 or 1970. A friend of Peter – named Steve - was also destined to experience the high-strangeness. Peter told me that he and Steve "...were big movie fans and great fans of Roman Polanski's movies. We loved *The Fearless Vampire Hunters* and *Rosemary's Baby*. This night we were listening to the soundtrack to *Rosemary's Baby* by Krzysztof Komeda. As one of the tracks ended, I noticed a kind of change in the air, a shift, a weird shift. It was a change in the mood of the place. And then the black mass track came on. Then things really changed. It seemed to me like we were in my living room, but also some place else."

Then, something even more disturbing happened, as Peter explained, "There was a knock at the door. I remember the knock at the front door. I don't recall if it was me or Steve who said: 'Better let them in.' I opened it; we admitted two men into the house. After what we had just seen, in retrospect this seems amazing. They were pretty much as you described on *Coast to Coast*: they were dressed in square, Eisenhower-era cop clothing, or FBI clothing – which in 1969, 1970 was not that unusual. They came in and sat on the couch. They were pale and sickly; their clothes hung real loose and they looked as though they might expire at any moment. They appeared to have either trouble breathing, or trouble even *being*. I don't believe they said a thing. If they did, it has disappeared from memory. Very odd, indeed."

Peter's complete, and *much* longer, story can be found in my 2015 book, *Men in Black: Personal Stories and Eerie Encounters*. And, I have to say, it is one of the strangest MIB-themed cases I have ever come across.

I also had the opportunity to extensively interview a woman named Alison (who I first spoke with in 2011), who had a very bizarre experience in a motel in Orange County, California, one year earlier, in 2010. At the time, Alison was in California on business. Her work was, and still is, in the field of acupuncture. It was an hour or so before the witching-hour when a loud knock on the door of her room had Alison in a sudden state of fear. This was hardly surprising. It was late at night, the sky was dark, and she was in bed watching none other than *Rosemary's Baby*. She lay frozen for a few seconds, then got up and tiptoed to the door and looked through the spy-hole. She could see two young boys, dressed in black hoodies and black jeans. As if sensing that Alison was watching them, one asked, in a raised voice, if they could use the telephone. Particularly disturbing, both "boys" held their heads low, clearly making sure that Alison would not see their faces too well.

In mere moments, Alison got to see the faces of the boys: they both had completely black eyes. No prizes for guessing who they were: the Black-Eyed Children, of course.

In early November 2016 I received in the mail a copy of Linda Godfrey's new book, *Monsters Among Us*. In her book, Linda tells the story of a man named Paul who, in October 2012, shared with her an intriguing story. Paul stated that at the time he was twenty-one and, one week before his traumatic encounter, he read Ira Levin's *Rosemary's Baby* novel. That Paul thought it relevant to even raise that issue is interesting. According to Paul, he was asleep in the front bedroom of his girlfriend's house when he quickly woke up – to a smoke-like odor. In an instant, Paul saw at the foot of the bed what he described as a "Dogman." He added: "It was very dark in color, like a German shepherd without the saddle colors but more black, and the presentiment of its intellect was very scary."

As if Paul's experience wasn't enough, only two days after Linda's book arrived, I received a Facebook message from a woman named Natalie, who had her own *Rosemary's Baby*-themed account to relate. Natalie told me that as someone who is a big fan of horror movies, she was very pleased to receive at Christmas 2015, a DVD copy of *Rosemary's Baby*, which she watched on the night of December 26 in her Austin, Texas apartment.

During the early hours of the 27th, Natalie experienced a classic example of sleep-paralysis. Looming over her – as she lay in bed and unable to move – was a darkly-hooded, man-sized figure with a pale face. And holding a sickle, no less. Natalie says she was given a "message"-like prophecy concerning the detonation of a nuclear weapon in early 2017 in a major city. Where, exactly, or even in which country, Natalie wasn't told. She "sensed" it would be a devastating, localized event – with millions dead – but which was quickly destined to spiral out of control and ignite World War Three, with *billions* dead. That Natalie shared her story with me less than 48-hours after Linda's book arrived is something I find intriguing. Of course, the fact that prophecies – and the supernatural entities that so often provide them – are notoriously unreliable suggests we should not be too concerned, despite the dangerous state the world is in right now – a state which, in my opinion, only Bernie Sanders stood a chance of putting right. But, that's clearly not going to happen now. Anyway, I digress!

That's far from the end of the matter, however.

November 5, 2016

Tempting fate and a Man in Black arrives

Later that same night – November 4 - after getting back from a local bar with a few mates where we watched an English soccer game on the pub's big-screen, I decided to watch my own DVD edition of *Rosemary's Baby*, which I did around 11:00 p.m. The timing is important, as watching the movie meant that I would be immersed in it when midnight struck and we moved into the early hours of the 5th. There was, however, no weird atmosphere of the kind reported by Peter Beckman and his friend, Steve. No creepy, pale kids came knocking on the door. No Dogmen were seen prowling around my apartment block under a bright moon. But, to very slightly warp the words of Ray Bradbury, something wicked certainly this way came. It just took a little bit longer to arrive.

I woke up on Saturday, November 5 around 8:30 a.m. I got out of bed, pulled on some pajama pants and went into the living room. I peeked through the blinds, specifically to see what the weather was like, and got the shock of my life – well, one of them, at least. My apartment is on a second floor and the living room window affords me a view of not just my balcony, but of the grass-surrounded pathway that separates my block from the one immediately opposite me, too. No word of a lie, about forty or fifty feet away – and walking along the path in a strange, slow and shuffling fashion, and with his head down - was a Man in Black. Clearly elderly, he was wearing a wide-brimmed black fedora with a light-colored band around it, a dark suit, and a striped tie. Weirdest of all, his arms hung in an almost dog-begging style. For a moment, I simply stared, suitably dumbfounded. Then, I swung into action.

I raced into my office: I had to get a photo. In barely a few seconds I was back at the window, I flipped the blinds again and captured the MIB for posterity – right as he walked slowly below me. I then ran into the bedroom, grabbed a t-shirt and put on my Converse All-Stars and raced for the door. By the time I opened it, he was gone. I practically leaped the steps as I headed down to the path. Puzzled, I looked left and right, no use. I walked around the parking area, just in time to see him getting into a large car – not a black Cadillac, I should stress – with three other men - but not definitive MIB. In seconds, they drove away and were no more.

To say that, only hours after watching *Rosemary's Baby*, I encountered a Man in Black, made me think I had very possibly – and maybe even recklessly - tempted fate in just about one of the worst ways possible. I quickly phoned my agent, Lisa, who – at first, at least - thought that the whole experience might have been explainable in down to earth fashion. She asked, after I emailed her a copy of the photo to see for herself, could he simply have been an old guy who happened to be wearing a black suit and a black fedora? Well, yes, he could have; but, even so, the timing was incredible. I had my doubts this was all a coincidence. Grave doubts.

After all, on the previous day I had written about the connection between *Rosemary's Baby* and the Men in Black and the Black-Eyed Children. And, there he was, the very next morning – an MIB outside of my apartment, right at the very time I just happened to look out of the window, and less than nine hours after I finished watching *Rosemary's Baby*. Plus, his clothing was very out of date and extremely conspicuous. On top of that, he certainly was not a resident of the apartments where I live; I knew that much for sure. When I told Lisa about the most important factors of all - the movie connection, and the timing of the encounter - a chill went through her bones.

It all amounted to an extremely uncanny couple of days; ones that I still brood upon when the wind howls and the skies are dark. And, just maybe, when there are things roaming around outside, lurking in the shadows.

19

"SHE DIDN'T LOOK RIGHT"

November 7, 2016

"About ten years ago, I had a possible Women in Black experience"

With my temporary excursion into the world of *Rosemary's Baby* now well out of the way, I phoned John Frick around 8:00 p.m. to get the lowdown on the MIB-related experiences of both John and his brother, Tim. As John told me, it was during their first trip to Point Pleasant in August 2002 that they got to see just how weird the town can really be. They checked into a hotel and, around 10:30 p.m. decided to do a bit of late night investigating:

> We saw some sort of a shooting star shooting across the sky in the direction of the Ohio River. It had a bright white head and flames shooting out of the back. It was pretty amazing. I thought it was just a fireball. But, in retrospect, we thought: what are the odds we would see one of the most impressive shooting stars we've ever seen in our lives? We'd never seen anything like that before. We had been out in the TNT area for ten or fifteen minutes and we saw this; it was pretty cool. Then, the next day, we were wandering around Point Pleasant for the first time in the day and we met with Carolin Harris, who helped with the Mothman Festival.

She said: "About an hour before you guys got here, there were six car-loads of U.S. Marshalls in town."

She said one of the guys came into the steakhouse where she was. He said something like, "They're not in here." And then he left. I joked that it may have been Men in Black looking for what we had seen the night before. We were joking about it, though. But, again in retrospect, we wondered what was going on. We kind of concluded that it was government officials in the area.

Flash forward to October of 2003, and we were visiting a friend in Point Pleasant. We left her house and we're driving to the TNT area. I said to my brother, Tim: "Maybe they really were the Men in Black."

I'm guessing within about fifteen or twenty minutes of me saying that, way off in the distance, from right to left, we saw a helicopter fly from West Virginia into Ohio. It circled the power plant and came back across the river into West Virginia. By this time we were getting up to the armory and we see the helicopter coming closer and lower to the road. It's hard to estimate how low it was, but it could have been as low as seventy feet. It was way lower than helicopters are legally allowed to fly. It was directly over the road, coming straight towards us. It was really freaky. I kind of got startled by it. We drove under it, and as soon as we drove under it, it flew off to the right. I looked over and saw it flying away. My first impression was, these guys were coming after us. What do we do? He could outrun us. It was kind of scary, but I was relieved when it flew off. Tim kind of dismissed it at first as probably being a fluke thing. I was pretty sure that it was Men in Black-related. Especially because I just

> brought up the Men in Black seconds before we saw it. The timing was really bizarre. That was about it for that. That was Friday, October 10.
>
> On Sunday, we were at a cousin's retirement party. We were in a pavilion and a helicopter flew really close to the roof of the pavilion. One of my cousins said, "What's he going to do, land on the roof?" I got all excited. It looked dark green or black. I wasn't scared at all this time. In my mind, I felt for sure this was paranormal. It was like with Keel: they were messing with us now. We had just met Keel less than a month before he moved. We hung out and had dinner with him twice; hung out with him for at least six hours. Somebody said Keel's energy may have rubbed off on me and my brother, because a month after meeting Keel all this weird stuff was happening.

John had more to say…

> I have a friend named Judy that I met at a Pennsylvania Bigfoot meeting. I called her and asked if she was going to the Pennsylvania Bigfoot conference this Saturday. She said a very strange police officer had just shown up at her door, about an hour before I called. She said the guy had a really straight nose and a really funny laugh. The guy said he was investigating one of her neighbors. It blew my mind, because just twenty minutes before I called her, I was reading an email from a woman named Jerry that I met at the Mothman Festival, and in the email she was telling me about how, when she was a little girl, there was an FBI guy in their neighborhood investigating one of her neighbors. That was like the phenomenon was screwing with my head, like the kind of stuff that was happening with

Keel. So, all of these things were happening the month after we met Keel. It was pretty strange.

About ten years ago, I had a possible Women in Black experience. I was in CVS and there was a very strange-looking Asian woman in there. Her face looked incredibly strange. The pharmacist was looking at her, too. Her face was so hideous that I immediately looked away, but curiosity would get you. You had to look back and see if it was real; I did that a few times. It looked as if she had too big a mouth, but too low on her face; she didn't look right. I walked out and said to my brother, "I think there's a Woman in Black in CVS; I'm going to wait for her to come out." Anyway, after two or three minutes, she comes walking out. She was wearing a skirt and high-heels. She walked across the sidewalk, and as soon as she hit the parking lot, she broke into a full run – across the parking lot. In high heels. As she passed me, she started laughing and got into the vehicle sitting next to ours. It had a guy with black hair in it. The vehicle pulled out and circled our vehicle; they looked at us, and circled around and left. It was weird.

John continued to me:

My other possible Woman in Black was when I was sitting in the lobby of a hotel; it was seven or eight years ago in Point Pleasant. It was around 10:30 at night. The rooms were kind of warm and we were letting the air-conditioning kick in. So, I came down and sat in the lobby to cool off some. I drifted off to sleep and I woke up when the front door rattled; they locked the front door after eleven. There was a woman there.

> I walked up to the door and I said, "Can I help you?" She said she wanted to see about possibly getting a room. I let her in and we were walking to the desk. And as I was talking to her I noticed that her eyes seemed to be half-rolled up into her head. That's how they were the whole time I was talking with her. She had reddish-brown hair and I thought the experience was weird, but I wasn't frightened or disturbed. But her eyes were really strange. Then, one of those synchronistic things about it, she started complaining about the traffic on route 35. Me and my brother had been on that road earlier that day. And I was talking about how bad the truck traffic is there. Then, having this woman come in and complain about the trucks was sort of synchronistic. Like they were messing with my head again. The next day, me and Tim met with a friend, Bobby, she's a woman who lives around Point Pleasant. We were going to check out some different areas and I was completely exhausted that day. I'm usually all pumped up at Point Pleasant; full of energy. I was thinking that she maybe psychically attacked me, the woman in the lobby the night before.

John and Tim's research into Mothman, the Women in Black and the MIB continues.

November 14, 2016

The "M.I.B. Grave"

On the night of November 8, 2016, Denise Rector flew into Dallas-Fort Worth International Airport, Texas to spend seven days with me. It was a week dominated by a fun road-trip that took us from Arlington to Austin, from Austin to San Antonio, and back to Austin. One of the things I did was to take Denise to see what I call "The M.I.B. Grave." You may well wonder what it is! No, it's not the final resting place of a pale-faced, fedora-wearing Man in Black, but you might be forgiven for thinking that's *exactly* what it is.

Western Heights Cemetery is located at 1617 Fort Worth Avenue, Dallas, Texas. It's a tiny, blink-and-you'll-miss-it-type of resting place for the dead. That is sandwiched between two small, old roads. Not particularly well looked after, the cemetery has most definitely seen better days. It has, however, a notable claim to fame. It's the final resting place of Clyde Barrow – as in Bonnie and Clyde, the infamous gangsters who lived and died by the bullet. Actually, by a hell of a lot of bullets. As for Bonnie, she's buried in Dallas' Crow Hill Memorial Park Cemetery, which is a fairly short drive from Clyde's grave. Buried next to Clyde are the remains of his brother, Marvin, who also came to a bloody end.

As we drove to the cemetery, I explained to Denise why I call it "The M.I.B Grave." The answer is very simple: Marvin's full name was *M*arvin *I*van *B*arrow: MIB. And, there is a small headstone at the grave which reads, in capital letters, "M.I.B." When we arrived, we found that –as is normally the case – the cemetery was locked with a large, metal lock. Fortunately, we were able to pull the gates wide enough to allow us to squeeze through the gap – something I am forced to do every time I take someone to see the grave. It's either that or jump the fence – when the cops aren't around to see.

As we stood around the graves I told Denise how, on several occasions, odd synchronicities had occurred in relation to Marvin's grave and the real MIB. Back in the mid-1980s, for example, the Dallas police had responded to a number of calls – covering a four-day-long period – of a man in a black suit and a black hat who stood in the cemetery and stared at passers-by in a stone cold, eerie fashion. He was never questioned, or caught, and had a strange ability to disappear and reappear.

On top of that, on the morning of the first time I went there – in December 2012 – I had been working on a MIB-themed article that involved a Man in Black seen at a cemetery in Dorset, England in 1992. Imagine my surprise when, on reaching the cemetery only a few hours later, I stumbled on Marvin's grave. This all struck me as some kind of classic "Trickster"-type phenomenon at work. Fortunately, that deranged atmosphere was absent when Denise and I were wandering around the old gravestones. Or, maybe, *unfortunately* should be the case, as Denise is quite partial to all things supernaturally-based.

November 21, 2016

"And he did have a wide brim fedora on"

I met Lydia Samuel-Hanselman while lecturing on August 10, 2016 in Rochester, New York. I was there to give a presentation on the Men in Black for Cookie Stringfellow's Rochester UFO Group. In November, Lydia told me an intriguing story:

> It was a very, very strange experience. It happened this summer and there's a wonderful park that runs through east Rochester. It has falls and different things in. There's a good long path you can take from one side of the creek. And I was with my friend Sacha - he's from the Ukraine – and my little dog, Mateo. It was about eight-thirty in the evening, but it stays light here in the summer for a while. We were walking down the path and, all of a sudden, this guy appeared. It was hard to tell if he was behind us; he was just there. He was all dressed in black. He didn't have

> a white shirt on: he was all in black. And he did have a wide brim fedora on. He was very, very pale. He just sort of kept up with us. I thought: holy shit, we're going into the woods, and this spooky guy is there. He ended up veering off to the left a little bit and stood behind a couple of big trees. Sacha said, "You okay, brother?" The guy just nodded. He just disappeared: I didn't see him walk out of the woods or anything else.

I didn't find it all surprising when I learned that Lydia had multiple paranormal experiences throughout her life. That seems to be something typical to so many witnesses to the MIB. I added Lydia's case to what is an always-growing file.

20

"A STRANGE AND EARLY 'WAKE-UP CALL'"

December 16, 2016

A strange tale of a stranger file

In the early hours of the morning, I had what was clearly a classic, but brief, hypnopompic experience, I had a strong sense of a hostile and sly entity lurking near the bedroom door. It was extremely careful to stay in the shadows. In a rumbling voice it said just four words: "Neil Armstrong's FBI file." The next thing I knew, my phone alarm was ringing loudly, it was now 7:00 a.m. on a cold and cloudy Friday morning. But, those four, very curious words were at the absolute forefront of my mind from the very moment I got out of bed. Considering the circumstances, there was only one thing I could do. Namely, to check out the FBI's website, *The Vault* – to (a) see if such a file existed on the first man to walk on the surface of the Moon; and (b) determine if such a file had been declassified under the provisions of the Freedom of Information Act.

Sure enough, the news was good – on both counts. Yes, to my astonishment, a file *did* exist on Neil Armstrong. And, yes, it was available for handy download in PDF format from *The Vault.* I wasted no time in scrutinizing the file: what could be in it that was so important it warranted a strange, early-hours-based message from something supernatural? Incredibly, and near-unbelievably, as I scanned the file I found a story that had distinct MIB overtones attached to it. The world's most famous astronaut and the Men in Black? Yep, that's exactly how it looked.

The Armstrong file, I quickly found out, is not a long one. The FBI stated, in a covering note: "This release consists of 18 pages of FBI file references to Armstrong ranging from 1969 to 1985 relating primarily to requests for FBI name checks in consideration of executive appointment; no derogatory personnel information was found. Redactions were made primarily to protect the privacy of living persons."

One specific portion of the small file really stood out: on February 2, 1976, the FBI's Special Agent in Charge at the Cincinnati, Ohio office prepared a memo titled "Neil Armstrong – Information Concerning." It stated:

> On January 29, 1976, Detective [deleted], Lebanon, Ohio advised that he had been contacted by several people working at the Lebanon Town Hall, including his mother. [Deleted] stated that on the previous day, two individuals, one a male Negro, the second a white male, had appeared at the Town Hall asking numerous questions about Neil Armstrong, the former astronaut, what his address was, how many children he had, where his children went to school, and inquired if he frequently ate at the Golden Lamb Restaurant and other personal question.
>
> These two individuals also went to the Town Tax Map Department and viewed diagrams of subject's property and adjacent property and asked several questions in that department.
>
> According to [deleted] no one really questioned these two individuals, although they were suspicious and these

> individuals stated they had been in town just to take some photographs of the house, as they were tourists.
>
> [Deleted] advised the only description he was able to get was they were both fairly young and male, and well-dressed. The Negro appeared to have a type of necklace, which had a quarter moon with a star on it. No one observed what type of automobile or any other details about these individuals.
>
> [Deleted] advised that on the following day, [deleted] came to the Town Hall and was quite concerned as to these two individuals and made a statement to one of the town employees, "I didn't think they would go this far."
>
> [Deleted] stated he is providing the information because of the notoriety of the Armstrong family and the strange activities by these two 'tourists.'
>
> [Deleted] stated he could be contacted at number [deleted] and that if anyone were to contact his mother regarding this situation, he wished to be contacted first as his mother has a heart condition.

Having read the relevant section of the file a couple of times, it wasn't hard for me to notice the MIB-style aspects of the story. The two men were described as being "well-dressed," which strongly suggests they were wearing suits. The reference to the mysterious pair being "in town just to take some photographs of the house, as they were tourists," strongly and eerily echoed the actions of the so-called "phantom photographers" of MIB lore and legend and which John Keel investigated in the 1960s and 1970s.

The issue of the pair asking questions concerning "how many children" Armstrong had, and "where his children went to school,' very much reminded me of the actions of the MIB-linked

"Phantom Social Workers" and "Bogus Social Workers" which I wrote about in my *Women in Black* book. Also, I found it notable that the two men asked if Armstrong "frequently ate at the Golden Lamb Restaurant." As I also knew very well, both the Women in Black and the MIB are infamous for approaching people in diners and restaurants – although they rarely ever eat, or even seem to know how to eat. Such accounts proliferate in MIB history. There was also the matter of the "strange activities" of the weird pair – an issue that remained intriguingly unexplained in terms of what, exactly, the FBI meant by that.

And there was one final thing; something of potential great significance. It was the issue of the necklace which one of the two men wore around his neck.

As the FBI noted, the necklace "had a quarter moon with a star on it." As I knew, a crescent moon and a star are recognized as the symbol of Islam, even though the origins of the design are much older. I checked out the *Symbol Dictionary* to see what it had to say about all this:

"This emblem, commonly recognized as the symbol of the Islamic faith, has actually acquired its association to the faith by association, rather than intent. The star and crescent symbol itself is very ancient, dating back to early Sumerian civilization, where it was associated with the sun God and moon Goddess (one early appearance dates to 2100 BCE), and later, with Goddesses Tanit and even Diana."

The reference to the Islamic faith instantly made me think of the intriguing connections between the Men in Black and one of the most dangerous of all Islamic supernatural beings, the Djinn.

A strange and early "wake-up call," the FBI, a file on Neil Armstrong, two mysterious men, and potential links to Islam, to the Djinn, and to the MIB: What on Earth was going on? On one hand, I might have written all off this off in a down to earth fashion. On the other hand, however, given that I specifically received what I can only call a "supernatural message" in the dead of the night – a message implying I should check out the FBI's file on Neil Armstrong – made me conclude otherwise. It was, without doubt, a seriously strange day.

21

"THERE WAS A GUY TAKING PICTURES OF MY HOUSE"

December 19, 2016

"A tall thin man with fanatical dark eyes"

Steve Wills, a U.K.-based UFO investigator, sent me this particular note, which is part and parcel of a definitively strange can of worms:

> Hi Nick, This story has an MIB type element to it....not sure if you were aware of this one? Arthur Shuttlewood [a prominent English UFO researcher in the 1960s and 1970s] whilst researching local ghost folklore concerned with the Royal Oak pub in Corsley Heath, a village about four miles (6.4 km) west of Warminster [a Wiltshire, England town and a UFO hotspot in the 1960s]. According to Shuttlewood, the building had once been a monks' refectory that formed part of thirteenth-century Longleat Priory, on the site now occupied by Longleat House. It was haunted by the ghost of a monk in a brown habit. The Royal Oak's landlord had also told him that a "triangle of passages and tunnels" led from the pub to Cley Hill – a prominent landmark to the west of Warminster, with evidence of an Iron Age hill fort on its summit – and a nearby farmhouse at Whitbourne.

> Some weeks after the story was published in "a leading evening newspaper," Shuttlewood was contacted by the landlord who said he had been visited by "a tall thin man with fanatical dark eyes who claimed that the [tunnels] held the most precious secret or earthly relic of the Devil." After his request to visit the cellars was granted, the young man then asked about demolishing the cellar wall that was said to separate the inn from one of the tunnels. When the landlord refused permission, his mysterious visitor confided his firm belief that the talisman of the Devil, the golden ram of Satan, lay buried in the earthen walls of the tunnel, probably interred under Cley Hill itself.

I was indeed aware of the story. It's one that revolved around something called The Golden Ram of Satan. Steve Dewey and John Ries, a pair of UFO researchers who studied this affair closely, said, "The Golden Ram of Satan is, according to Shuttlewood, a covered talisman of the Devil, which lies somewhere in Europe, and which, when found, will grant its finder every wish."

Dewey and Ries continued, "Shuttlewood found out about the Golden Ram after investigating some local folklore pertaining to ghosts that had been seen at the Royal Oak public house in Corsley, a village about four miles west of Warminster." It was when Shuttlewood's story appeared in the press that the strange, MIB-like character referred to by Steve Wills appeared on the Royal Oak, acting in very peculiar fashion.

Precisely what all of this means was – and still is - anyone's guess. I couldn't help recalling, however, that the presence of a MIB-like character in relation to ancient mysteries is nothing new. As I knew all too well from my own studies of the MIB and WIB phenomena, they pop up in relation to such historical issues from time to time. For example, in 1996 an English freelance television producer, Michael Hartley, found himself subjected to MIB-type threats after researching the alleged final resting place of the legendary Robin Hood – supposedly near the Yorkshire, England village of Clifton. Similarly, my *Women in Black* book told the story of a man named Colin Perks. Late one night in 2000, he was visited by a WIB - and soon after by a red-eyed Mothman-like entity - after digging into the matter of the final resting place of the legendary King Arthur of times long gone. Perks suspected the Woman in Black and the winged fiend were one and the same: a dangerous shape-shifter.

January 10, 2017

"You have nine more years!"

It was on this day that one of the strangest – albeit brief – reports of the Women in Black kind came my way. This one is from Daryl Collins of Dallas, Texas:

> In early June of 1986 (don't recall the exact date), I was just coming from a discussion of my abduction memories and approaching the parking lot to drive home. A white-haired lady in dark or black clothes drove by in a dark or black car. As she approached me she slowed down, leaned out her window and yelled at me, "You have nine years until what you were made for! You have nine more years!" Then she sped up and drove away. I didn't catch the license number, make of car, or any other details. But I distinctly remember what she yelled, since I spent the next nine years haunted by what might be awaiting me in early June of 1995. When the date finally arrived, there were smells suggesting the things might have been in the

> house, but otherwise nothing happened. The literature is full of cases where predictions failed to come true; maybe they just change their minds.

As I read Collins' account, I couldn't fail to remember how - at the height of his Men in Black encounters in the early 1950s – Albert Bender found his Bridgeport, Connecticut home filled with overpowering odors of brimstone and sulfur. What goes around clearly comes around, too. In light of the above, I wasted no time in asking Collins a few questions.

January 19, 2017

"There followed some kind of physical examination"

On the particular matter of his abduction experience as a young boy, nine days later Collins shared with me his Woman in Black encounter. He told me:

> Approximately April or May 1945, an extremely bizarre encounter. I was playing in the backyard when I spotted a creature about my size, an elf perhaps, which communicated telepathically. It merged its body with mine and controlled my movements. We went out the back gate and down several streets to an area that at that time was still wild country. We found a hole in the ground and jumped in, falling a fair distance, stopped, and entered a strange room and walked down a corridor. The corridor led to a door that swung open, and the elf separated from me. There followed some kind of physical examination, then I was back in my yard quickly forgetting the whole incident.

Long ago I described the details to Budd Hopkins, and he said he had many similar cases.

January 20, 2017

"He was activating an implant that had just been put in place"

On the following afternoon, there was this from Collins:

> In 1948, my parents and I were driving at night on a deserted rural highway. An object came over and lifted the car into a big round room. A door opened and three creatures, with clawed webbed toes and stubs for fingers, took each of us in different directions. I followed one down a long dark corridor to a brightly lit room with an examining table and a gray with long bony fingers. He stuck a sharp instrument into my stomach, then a different instrument far up my right nostril. When he pulled it out, my nose bled a little and was sore for hours. Then he took an instrument like a black cone and pressed the point to my forehead, producing a strange vibrating sensation. Apparently he was activating an implant that had just been put in place. He said, telepathically, "It will be all right from now on." Abruptly we were all back in the car and the whole episode was forgotten.
>
> I don't remember the date, but once I was confronted with a large insectoid that said, "I want you to devote your whole life to me!" I didn't understand and asked what that meant. I don't remember what the creature said, but I replied emphatically, "No, I won't!" This evidently didn't go over very well. In January 1950, I was taken at night

> from my home by skeletal creatures that said they were going to kill me. I begged for my life, reminding them of the time they had said, "It will be all right from now on." So instead they put me on the table and performed extensive procedures. Finally they abandoned me and all the memories were lost for very many years. As far as I can tell, I was never abducted again. When I first encountered "flying saucers" in the newspapers that March, I immediately adopted the subject as the center of my life, but it never occurred to me to wonder why, nor to suspect it had anything to do with me.

In view of Collins' experiences of the UFO variety as a young child – in 1945 and 1948 – it seemed to me hardly surprising that he received a chilling message from a Woman in Black in the 1980s. I knew only too well that both the MIB and the Women in Black often intrude upon the lives of those who have had profound UFO encounters – and sometimes years later. And the experience is seldom, if ever, positive in nature. It certainly wasn't for Daryl.

February 1, 2017

"His face was pale, like super-white"

Alissa Plakaros is a friend who I see at conferences, now and again – and most recently at the September 2016 Mothman Festival. On January 30, 2017, Alissa contacted me to say that she had just undergone a strange MIB-themed encounter in the Florida town of Mims. Well, I had to follow-up on that! Two days later I caught up with Alissa by phone. She told me, "It was Monday. This was in town; we had gone to a store. Mom and I were joking, as we saw this guy who looked like a Man in Black. He had the black suit on and was very well-dressed, with sunglasses. His face was pale, like super-white. He threw me, as he walked back to his car – which was a newer car and black - and was looking at me when I said to mom he looked like a Man in Black. He wasn't close to me, but it was weird; like he could hear what I was saying about him being a Man in Black. When I said that he turned and looked at me."

Alissa then proceeded to tell me something very interesting. On the night of the encounter with the pale-faced MIB, she had a UFO sighting; a cluster of lights in the sky which, said Alissa, "were changing to all different colors – blue and red - and moving really fast."

February 21, 2017

"There are definitely people who have been recruited"

For more than a decade, Lisa Hagan has been my literary agent, and a fine job she has done for me. She is also a good friend. Not only that: in the last few years Lisa has experienced a few very strange things, those things that are all too common in certain MIB-based encounters. I say "certain" ones, because Lisa's encounters seem to be specifically with the government type of MIB, rather than with the clearly paranormal kind of Men in Black. Remember: John Keel was certain there were *two* groups of MIB. Maybe there are even more factions.

It was on February 21 that I got around to interviewing Lisa about her many and varied experiences, all of which, when combined, point to the fact that she has been under close surveillance for a very long time. We'll begin with my 2005 book, *Body*

Snatchers in the Desert – a controversial book which was focused on a non-alien theory for what happened near Roswell, New Mexico, way back in July 1947.

Virtually all of the research was completed for *Body Snatchers in the Desert* by the end of 2003, and the first seven or eight months of 2004 were spent writing the manuscript for Simon & Schuster. Only days after I mailed a copy of the Word document to the publisher – yes, back then, publishing houses still wanted a paper copy, as well an emailed attachment – it vanished, overnight, from the offices of S&S in New York. Josh Martino and Patrick Huyghe, who both worked on the editing of the book, were perplexed – and more than a little disturbed and concerned, too. Jokes about the manuscript probably being held in the underground vaults of the NSA may not have too far from the truth, Patrick suggested. And not in entirely tongue-in-cheek fashion.

I have to confess that I had totally forgotten about this odd event – that is until I decided to interview Lisa for this book. That was when something came flooding back into my mind; something Lisa had told me years ago and which concerned links between publishing houses and the government / intelligence community. And, so, this was the ideal time to finally get the story nailed. As Lisa told me on the morning of the 21st, it's practically par for the course, when it comes to books on controversial subjects that might prove to be bad for your health, for government agencies to sit up and take notice – and even before the publication date:

> If I want to do anything that is going to blow the whistle on something that the government doesn't want out there publicly, the government can shut that down – at the publishing house. They have people on the inside. It took me a long time, after being rejected on certain projects,

to finally realize that what was happening to me was the same thing that had happened to my mother. Shut down by publishers on the stories the government didn't want out there.

I still work with these publishers – they do their job; I do my job. I do know that one of the publishers I worked with in New York – this was back when all the manuscripts were printed – they had certain people who read the manuscripts *before* the editors got to them. I don't know which way it goes: maybe, they were recruited by an agency when they were already working for the company. Or, they infiltrated it. But, I do know of people in the publishing industry who have been asked to be informants – asked by people in the government. There are definitely people who have been recruited.

MIB in the book-publishing industry? Yep, that's exactly how it looked. And as someone whose books have made the front pages of *The New York Times, USA Today, The New York Times Book Review,* and *The Los Angeles Times,* Lisa should certainly know what she is talking about. We then got into something of an equally sinister – but very different – fashion. Someone, Lisa told me, had been prowling around her property under cover of darkness:

It's very dark where I live and there are only a few people here. It's very heavily wooded and there's a lot of farmland. And it's all dirt road, so there's not a lot of traffic. So far, it has happened to me at least three or four times. My house is a long bungalow, and my bedroom is on the end, near the carport. Periodically, I'll be lying there reading with the light on. It's always been between 10:00 and 11:00 at night, and I'll hear my vehicle trunk open. Then I hear

> it close. I do have a picture of a handprint in the dust on my trunk that I took after it happened one night. I kept the picture in case I wanted to do anything with it, which I did not.
>
> If I had wanted to, I could have crept up in my pajamas and intervened, but what would I do? It made me mad, though. Nothing was ever taken from the trunk. But, I did check if anything had been put in there. Those bugs are small though, so who knows?

Then, there was the matter of Lisa's encounter with what sounded just like one of John Keel's "phantom photographers." Lisa told me the strange story:

> It was a Sunday and I was in the kitchen making something, standing at my kitchen sink. And, I saw a flash to the left. In my yard. I looked over and there was a little white Honda. An Accord, I think; two people. There was a guy taking pictures of the side of my house. And then they drove right in front of the house and took a picture – right where I was standing in front of the kitchen window. They were only about two car-lengths from the window. They went down passed my house and then came back. And then they drove off. That made me nervous and my anxiety went up. My intuition, my alarm bells, went off.

And, on similar territory, there was this from Lisa:

> The other story is that my aunt and I had seen a UFO on our farm. This was Christmas night, 2015. And, on the 26th she left. I called and asked friends if they wanted to come over, to look for UFOs. They were like, "No, we're already

> in bed." But, another friend wanted to come over and see if we might see anything. We were standing by the trunk of my vehicle – the trunk someone had been in and out of – as we had the iPads and the binoculars resting on it. It was a beautiful night and we were talking and looking around. As my friend left, he said that when he was maybe ten cars away from me, around a row of trees here, there was a black, unmarked SUV with a lot of antenna on it. Not long after that, the same friend was driving on our main road – the road that everyone lives off – and there were two of those SUVs, with men outside with binoculars, overlooking our farm.

Then, there was what may have been the most sinister account of all. Lisa told me:

> One other thing that happened: I have been working on a book on Gary DeVore; he's dead now, but worked with the CIA. And oddly enough, my author is dead now, too. When he would send things to me, like his manuscripts and pictures, I would go to the mailbox and some of it was opened and there was nothing in the packages. Other times just one thing would be missing. Or, the packages would have been opened, but everything was in there.

I knew only too well the story of Gary Devore. In September 2015, the U.K.'s *Daily Mail* newspaper reported: "When the skeletal remains of Hollywood screenwriter Gary Devore were found strapped into his Ford Explorer submerged beneath the California Aqueduct in 1998 it brought an end to one of America's most high profile missing person cases. The fact that Devore was on his way to deliver a film script that promised to explain the 'real reason' why the U.S. invaded Panama, has long given rise to a slew of conspiracies surrounding the nature of his 'accidental' death."

The *Daily Mail* continued: "It didn't help that Devore's hands were missing from the crash scene, along with the script, and that investigators could offer no plausible explanation as to how a car could leave the highway and end up in the position it was found a year after he disappeared."

As the *Daily Mail* also noted: "...Devore was working with the CIA in Panama and even a White House source concedes his mysterious death bears all the hallmarks of a cover-up."

It's a mystery that still persists. And the mystery of Lisa's MIB encounters persists, too.

22

"THEY WOULD POINT THEIR BONY FINGERS"

March 2, 2017

MIB, the Philadelphia Experiment, and a "gift" on the doorstep

Something extremely strange happened across March 2 and March 4. Something that convinced me, even more, that I was being watched by strange forces – some of them human and others most definitely not human at all. It all revolved around a highly notorious affair that has become known as the "Philadelphia Experiment" and placed the MIB right in the heart of the controversy.

Back in 1955, Morris K. Jessup's book, *The Case for the UFO*, was published. It was a book that delved deeply into two key issues: (a) the theoretical power-source of UFOs, and (b) the utilization of the universal gravitational field as a form of energy. Not long after the publication of the book, Jessup became the recipient of a series of extremely strange missives from a certain Carlos Miquel Allende of Pennsylvania. In his correspondence, Allende commented on Jessup's theories, and gave details of an alleged secret experiment conducted by the U.S. Navy in the Philadelphia Naval Yard in October 1943. Thus was born the highly controversial saga of what has become known as the Philadelphia Experiment.

According to Allende's incredible tale, during the experiment a warship was rendered optically invisible and teleported to – and then back from – Norfolk, Virginia in a few minutes, the incredible feat supposedly having supposedly been accomplished by applying Albert Einstein's never-completed Unified Field theory. Allende elaborated that the ship used in the experiment was the DE 173 *USS Eldridge*; and, moreover, that he, Allende, had actually witnessed one of the attempts to render both the ship and its crew invisible from his position out at sea on-board a steamer called the *SS Andrew Furuseth.*

If Allende was telling the truth, then the Navy had not only begun to grasp the nature of invisibility, but it had also stumbled upon the secret of teleportation of the type demonstrated – years later, in fictional, on-screen format – in *Star Trek* and *The Fly.* On these very matters, Allende made the disturbing claim that not only did the experiment render many of the crew-members as mad as hatters, but some, he said, even vanished – literally – from the ship while the test was at its height, never to be seen again. Others reportedly suffered horrific and agonizing deaths. There are even tales suggesting some of the crew were flung into the future, specifically to the heart of the 1980s.

Of course, as students of this very weird affair will know, the tale of Allende and the vanishing ship (or non-vanishing ship, depending on your perspective!) has been denounced as much as it has been championed. But, few are aware of the U.S. Navy's official stance on the matter. Many assume – quite incorrectly – that the Navy's position is that nothing whatsoever occurred at all. But their assumptions are wrong.

Contrary to what you might think, the Navy does believe the story has a basis in fact – albeit of a far more down to earth nature. While there is certainly no official endorsement of the stories that the *USS Eldridge* was rendered invisible in October

1943, that it was teleported from one locale to another and then back again, or that crew-members were injured, killed, or outright vanished into oblivion, the Navy does admit that, in all likelihood, the story has a basis in something of real, secret significance.

The Navy's current position reads as follows: "Personnel at the Fourth Naval District believe that the questions surrounding the so-called 'Philadelphia Experiment' arise from quite routine research which occurred during World War II at the Philadelphia Naval Shipyard. Until recently, it was believed that the foundation for the apocryphal stories arose from degaussing experiments which have the effect of making a ship undetectable or 'invisible' to magnetic mines."

Degaussing, in simple terms, is a process in which a system of electrical cables is installed around the circumference of a ship's hull, running from bow to stern on both sides. An electrical current is then passed through these cables to cancel out the ship's magnetic field.

"Degaussing equipment," says the Navy, "was installed in the hull of ships and could be turned on whenever the ship was in waters that might contain magnetic mines, usually shallow waters in combat areas. It could be said that degaussing, correctly done, makes a ship 'invisible' to the sensors of magnetic mines, but the ship remains visible to the human eye, radar, and underwater listening devices."

Just to confuse things, however, the Navy has offered a further theory, too, to explain what might lie at the heart of the story:

"Another likely genesis of the bizarre stories about levitation, teleportation and effects on human crewmembers might be attributed to experiments with the generating plant of a destroyer, the *USS Timmerman.* In the 1950's this ship was part of an experiment to test the effects of a small, high frequency generator providing 1000hz, instead of the standard 400hz. The higher frequency generator produced corona discharges, and other well-known phenomena associated with high frequency generators. None of the crew suffered effects from the experiment."

That the Navy first denounced the Philadelphia Experiment as having any basis in reality, but today is seemingly happy to offer no less than *two* theories to explain what might have been behind the legend – involving verifiable, secret projects – has inevitably raised suspicions that we are still not being told the full story of what really occurred all those years ago at that mysterious naval yard, regardless of how one views the strange saga of Carlos Allende.

With that all said, it's now time for me to get to the events of March 2. As someone who writes books, gives lectures, and posts articles here and there, I get a lot of feedback. For the most part, it comes via emails, phone-calls and Facebook messages, from people who want to share their experiences, ask questions, or offer their opinions on the things I have written about. Occasionally, however, I'll find myself on the receiving end of a communication (or several) from someone claiming an "Insider"-type background. A whistle-blower, in other words. There have been a few occasions when my path has crossed with those of enigmatic characters whose backgrounds did indeed place them in the world of certain, covert activities - all connected to UFOs, in one way or another. This occurred most graphically when I was researching and writing my books, *The Roswell UFO Conspiracy*, *On the Trail of the Saucer Spies*, *Body Snatchers in the Desert*, and *Final Events.*

On the 2nd, it happened again. In this case, it came in the form of an "Unknown Caller" phone-call in the early afternoon. The elderly man at the other end of the line said that he had read my books *Men in Black* and *Women in Black* and wanted to share something relevant. It all revolved, he said, around the Philadelphia Experiment of 1943 and the MIB. This was certainly a new one on me. And I thought I had heard it all! I clearly had not.

Like me, he had no time for the teleportation- or time-travel-based scenarios. But, what he did believe (or, rather, claimed to know) was certainly just as controversial. Perhaps even more so. He said that in the early 1980s he was given the opportunity to read a particular batch of classified U.S. Navy files that told the "real" story of the Philadelphia Experiment. He was extremely cagey on verifiable facts (no surprises there, I have to say...), but maintained that the experiment had a bearing on the aforementioned MIB and Women in Black. I asked, "In what way?"

His reply was that the experiment - the precise nature of which he would not comment on - created what he called "a shift." It was a shift that allowed those aboard the ship to see certain things that, in a normal state, they, you and I would not be able to see. But, which are supposedly around us all of the time. We're talking about the MIB, WIB, and even what have become popularly known in recent years as the Shadow People, one-dimensional entities that are considered highly dangerous.

So the tale went, the stories of crewmembers vanishing, or becoming fused into the metal of the ship itself, were very wide of the mark. They were, I was assured, wildly distorted accounts of the crew seeing not sailors, but encountering MIB, WIB and Shadow People walking through walls, becoming invisible, and then reappearing. It didn't take long, though, before the "truth" of the matter became a tale of vanishing and reappearing sailors. The man claimed - even more controversially - that this particular event at sea marked the first moment when the U.S. Government became "aware" of the Men in Black phenomenon.

The man stressed that he used the word "aware" for a very good reason. He wanted to make it clear that the government's awareness did not mean they fully understood the nature of the MIB (and related) phenomenon. And, then, he stressed *yet again* that awareness and understanding should not be confused. *Yes, I got it, jeez.* Interestingly, he also said that the stories of some of the crew going insane were absolutely true. He explained that this was caused by the strange ability of some of the men to randomly see the MIB, the WIB, and the Shadow People for up to at least several years later. True or not, I could understand why people might well flip their collective lids under such circumstances. Imagine being endlessly faced with ghoulish, pale creatures in black swarming all around you, all the time, and in somewhat ethereal form. And your friends and family are completely oblivious to what's going on around them. That would surely be enough to send anyone completely off the rails.

Creepiest of all, he said that when the creatures realized they could be seen by the sailors, they would respond by endlessly tormenting them. They would point their boney fingers at the men and grin in manic style. Others would "dance" around them, in wild, crazed fashion, while wailing and howling. It was a definitive bedlam, one from which there was no escape whatsoever. A *Danse Macabre,* one might say. Minds were lost, destroyed and never recovered. It reminded me very much of Daryl Collins' 1986 experience with a mad, shrieking Woman in Black, the menacing words of which tormented Daryl for years.

Having listened to all of that, I felt like I needed a drink. Was 3:00 p.m. on a weekday too early for hard liquor mixed with chilled Coca-Cola and ice? No, not today, it certainly wasn't. Things weren't at an end, though.

March 4, 2017

A visitor in the night

As interesting as the story admittedly was, I quite naturally asked the man for something to back up his extraordinary claims. He told me to keep a look out for something that would be "arriving" in a couple of days. The phone then clicked. He had hung up. I thought, how can something be coming to my apartment when he didn't even have my address? Two days later, on March 4, I found out. Apparently, he, or someone associated with him, *did* have my address. Around noon, I went down to get the mail. As I opened my apartment door, I saw a yellow envelope sitting on my "Welcome" mat.

What was particularly odd about the envelope was its complete lack of stamps or address; the entire envelope was blank. It wasn't even sealed; the flap had just been pushed inside. I looked around but didn't see any unfamiliar faces wandering around. That was not surprising, it could have been placed there any time after I last returned to the apartment, which was around 10:00 p.m. on the previous night, after I had helped a friend haul a new recliner into his second-floor apartment, one block down from mine. Logically, though, I figured it was probably delivered in the early hours of the morning, when there would have been hardly anyone awake and in a position to see the person who made the stealthy delivery.

So I took the envelope inside and sat down on the couch. I opened it and could see what looked like an old book. That's exactly what it was. I took it out. What I was holding in my hands was an original, 1950 edition of a book published by the Bureau of Ships, U.S. Navy Department. Its title was *Microwave Techniques*. It had been prepared by the Radiation Laboratory of the Massachusetts

Institute of Technology. On the "Acknowledgments" page, it said, "Microwave Techniques is a Bureau of Ships edition of the report T-13, Microwave Technique as of May 1943, issued by the Radiation Laboratory." It was signed "D.H. Clark, Rear Admiral, U.S. Navy, Chief, Bureau of Ships."

When, back in the 1980s and the early-to-mid-1990s, I used the Freedom of Information to a significant degree, I was used to receiving government-, military-, and intelligence-based documents through the mail. But, this was very different. Via FOIA you would always get photocopied material. And, of course, it would be delivered in conventional fashion. On this occasion, though, I was the recipient of an *original* military document. Plus, it wasn't delivered by the mail, by UPS, or FedEx.

On this occasion, someone had climbed the stairs to my apartment and had very quietly placed it right outside of my door (in the dead of night, I concluded), knowing that I would soon find it. Also, there was the not insignificant fact that the "Acknowledgments" section of the *Microwave Techniques* document referenced 1943 – the year of the Philadelphia Experiment. The report didn't mention the legendary experiment, but there was no doubt in my mind that it came from the old man on the phone two days earlier. Or, at the very least, from an associate of his.

After perusing the book for an hour or so, I went outside and scanned the apartments. The white, semi-feral cat who I feed every day with a large handful of Temptations was lounging around at the foot of the steps and meowed as he / she saw me. Manuel, the maintenance guy, was busy as always – fixing someone's air-con unit. Everything was as normal as it ever was. Or, rather, it *seemed* to be. Beneath the veneer of normality, though, I detected a sense of something else. Of someone who had tainted my home and as I had slept the night before. Of dark machinations and of an unclear, masked agenda.

The affair didn't disturb me, but it definitely made me recognize there were strange and manipulative characters out there. They were teasing me with tales of MIB, WIB, and Shadow People, creeping around my apartment block when darkness fell, and dropping off decades-old military documents – almost literally into my lap. Of course, had they tried to make their way *inside,* they would have found themselves filled with considerable amounts of lead. I live in Texas. And guess what? I know the gun laws in the Lone Star State *extremely* well.

23

"THERE WAS AN OLD, BLACK LINCOLN CAR COMING TOWARDS US"

March 13, 2017

Men in Black and their black cars

As I know only too well, and as a result of years of looking into the MIB mystery, it isn't just the MIB that are weird. Their cars are, too. Just like the dark-suited characters that drive them, their vehicles also display evidence of deep and bizarre high-strangeness, including having the ability to vanish into complete oblivion. Even to be able to fly! They are definitely justified in being called CIBs: Cars in Black.

Dr. Josef Allen Hynek, a well-renowned UFO authority who died in 1986, was provided with the details of a fantastically-strange MIB encounter that occurred in a small Minnesota town in late 1975, and that falls firmly into just such a category. Ironically, no UFO was seen on this particular occasion, but the chief witness was harassed by the driver of a large, black Cadillac on a particular stretch of highway, and who nearly forced the man into an adjacent ditch. The irate man quickly righted his vehicle and headed off in hot-pursuit, only to see the black Cadillac lift into the air and, quite literally, disappear in the blink of any eye.

Long-time UFO authority, Jenny Randles, the author of *The Truth Behind Men in Black*, investigated an extremely similar case that occurred in Britain in August of 1981. In this particular incident, the witness, one Jim Wilson, had seen an unidentified – but not overly-fantastic – light in the sky, and was later blessed with a visit from a pair of suit-wearing characters flashing I.D. cards that demonstrated they came from the British Ministry of Defense. The two suggested to the man that he had merely viewed a Russian satellite – specifically Cosmos 408 – and that he should forget all about the experience.

That would indeed have been the end of things, were it not for the fact that the witness found to his concern that, on a number of occasions and shortly after the visit occurred, his home seemed to be under some form of surveillance. By two men sitting in a black *Jaguar* car – which is the preferred mode of transport in most British Men in Black cases. The police were called, and, across the course of several nights, stealthy checks of the immediate vicinity were made. After seeing the car parked outside the man's home on several occasions, and then managing to get a good look at his license plate – which the police were quickly able to confirm as being totally bogus – they carefully closed in, with the intention of speaking with the pair of MIB and finding out the nature of their game. Unfortunately, they never got the chance to do so.

As two uniformed officers approached the vehicle and prepared to knock on one of the windows, the black *Jaguar* melted away into nothingness. There was, not surprisingly, a deep reluctance on the part of the officers to prepare any kind of written report alluding to such an event in the station log-book. Just like the Men in Black themselves, even their vehicles are seemingly able to perform disappearing acts of a near-ghost-like nature.

There is a very good reason I bring the issue of "Cars in Black" to your attention. While going through a bunch of old files in early March 2017, I stumbled upon a few scribbled notes from years earlier that concerned good fiend Tracie Austin and the MIB. As you'll recall, Tracie shared with me her Black-Eyed

Children cases in late 2016. On finding those notes, though, I called Tracie on March 13 and asked her if she could expand on this other story. Time to get it all down, once and for all. It all revolved around Tracie's personal encounter with a CIB, back in the late 1990s. She said:

> I remember when you and I went to the *UFO Magazine* conference in September 1999, in Leeds [England]. I got the train there, you drove there, and then you drove me home, afterwards. Well, it wasn't long after that that this thing happened. I had moved up to Cheshire the week after I got back from the conference. While we were at the conference I had met Brigitte Barclay [a well-known alien abductee in the U.K.]. That was the first time I had met Brigitte. We just got talking and she happened to tell me of when she lived in the States she had Men in Black episodes. And how she was followed by black cars and black helicopters.
>
> So, I went to the conference, talked to Brigitte, and then was the house move. A few days after, a friend and I went to the movies and a Pizza Hut – it was a Sunday night; I remember that. We were driving from Cheshire into Staffordshire, for about fifteen miles. And Cheshire is full of country roads, as you know: Farmers' fields, countryside; that kind of thing.
>
> As we entered Staffordshire, I noticed there was an old, black Lincoln car coming towards us in the opposite direction. Although it was old, it looked new, but it was an old style, and it was totally out of place. And, coming from England, in Cheshire, that's just not the kind of car you see; an American Lincoln. That's what I thought; what's an American car doing here?

As the car approached us – coming the other way – I tried to see who the driver was, but I couldn't see anybody. I'm not saying the car was driving itself, but I just couldn't make out anyone. *No one.* I could see it had really shiny bumpers and there was no [license] plate, at all. And I noticed that with every car behind that Lincoln, it was like it was the leader of the pack. I could see the drivers of the cars behind were all kind of "switched off." It looked like an automatic mode. Like they were in a daze. It was very, very odd.

We got down the road and my friend said, "Did you see the way that vehicle was moving?"

I said, "No, what do you mean?"

He said, "The wheels were going around, but they weren't touching the ground."

It was just like it was hovering, but not high, or I would have known. I was too busy trying to see the driver. I just said, "Oh, my God." It was really odd that I had had that conversation with Brigitte only a week prior, and now, a week later, I was having my own Man in Black encounter. I don't believe in coincidence at all; I think everything is synchronicity.

April 8, 2017

"Nobody ever goes out there"

David Weatherly was due to be on *Coast to Coast AM* this night, speaking about his then-new book with Ross Allison, *Haunted Toys.* More correctly, in central time, at least, in the early hours of the 9th. There was, however, a problem: David doesn't have a landline; only an iPhone. And, *Coast to Coast AM* is absolute sticklers for demanding their guests use a landline – and with a cord, too. No problem. As I have a landline as well as a cell, I suggested that David come over to my place – he only lives about an hour or so away – and do the interview from my home. Which is exactly what happened. David came over around 10:00 p.m., and we had a good time hanging out, chatting, catching up, and all to the muted background of an Italian soccer game on the television; I have to get my regular fixes of soccer, even if they are in silence. It was while we were waiting for *Coast to Coast AM* to call – around 2:00 in the morning – that David shared with me his strange and personal encounters with MIB-type characters. Those encounters were, to be sure, surreal. Several of them, as you'll see, involved cars of the very mysterious type.

With David sat on one couch, and me on the other, I placed my digital recorder between us and hit the "record" button. David began…

> I've had several weird encounters in the area around Area 51, on trips out there. One particular time – this was 2014 - I was driving back from a conference in California. I hit Nevada pretty late and I realized that I would take a route that would take me past Rachel and Alamo. So, I would get the chance to stay there for a night. The only thing in the area is the Little A'Le'Inn [a UFO-themed bar]. I called ahead and checked with the guy who ran the little motel there in Alamo. Alamo is a good place to use as a base and sky-watch. He was fine and said to come along to the office and he would have a key there for me. It was a nice night, and with it being the desert, it was still warm. I rolled in – this little town is a speck, you know? There's nothing there; certainly nothing that stays open much past five o'clock.

I got in, and the motel was just one of those really old strip motels, probably from the sixties. I pulled off and parked my car and there were only a couple of other cars there in the parking-lot. I got my room-key and came out of the office and closed the door. I had my luggage – a roller – and my laptop, so my hands were full. My room was only a couple doors down from the office. But, as I'm moving towards my room, I see a guy leaning against a car in the parking lot. It was really odd: he didn't have a hat on, but he was wearing a black suit and a skinny, black tie – like an eighties tie. He was staring at me; and he wasn't there a moment before. It really stood out; really strange. He looks at me with this stern look and he says, "There's not going to be any coffee." It was so surreal; I was trying to process all the pieces. A couple of funny thing stand out in my mind. He had a military haircut, that black skinny tie, and his shoes were black dress shoes, but they were really shiny. Although he was trying to appear casual, he didn't seem relaxed. It was like it was forced. There was something very awkward about it. It was only a few steps to my door, so I put the key in and opened the door and I thought, I'm gonna see what this guy's deal is. I rolled my bag into the room and only had to take a step or two into the room, to toss my laptop onto the bed. And in those few seconds, when I turned around, and looked outside the door, he was gone.

As midnight approached, and as I listened carefully, there was more to come from David…

It's really curious because there's a handful of weird little incidents like that out at Alamo. There's another one that was probably a month later. There's another hotel there that has cabins; right down the street from the other hotel I was staying at. And it has a restaurant in their building; pretty decent food. I was there with a friend; we were staying in one of the cabins for the weekend. We were going to do some sky watching with night-vision goggles. We were in the restaurant for lunch and there were probably only three other people seated – and the staff.

Two guys enter in black suits. They have black sunglasses on, no hats. They walked in and they walked the whole restaurant, as if they were looking for someone. But they also stood out and they looked a little odd. They came back to the center of the restaurant where we were sitting, and these two guys stopped in the center and one of them, I noticed, had his phone up and was taking pictures of the inside of the restaurant, which, of course, was kind of curious. The other guy says, to no one in particular, "We don't want any fish in here." And they walked off.

One other time, in 2015, we were out sky watching near Area 51 late at night. And this van drove out around us, two or three times. Again, this was right around the cabins. There is a circular drive that goes around the cabins, and this van came around, old, a dark green in color. This guy pulled through and on about the third pass he put his window down. And he had rolled the passenger's side-window down, as we were on the passenger's side, too. You couldn't really see him very clearly, but he stopped and sort of made this proclamation and said, "People looking in the sky often see things they shouldn't see." And then he drove away. That was the last we ever saw of him.

The weirdest story of all was still to come…

Although Wikipedia states that the location of Nevada's Lovelock Cave is "restricted," it's actually very easy to find: it's situated south of the town of Lovelock, Pershing County. It's a sizeable, shadowy cave – around 150 in length and 35 feet in width – and one which has a great deal of history and controversy attached to it. Cryptozoological controversy, one might well say. Excavations began in the early 20th century revealed that the cave was inhabited by local tribespeople for at least 4,000 years – and possibly even longer than that.

In 1911, a pair of miners – James Hart and David Pugh – hauled out from the cave tons of bat guano (shit, for the uninitiated). Their actions revealed something amazing; a large number of ancient artifacts that had been buried for an untold number of millennia. In the years and decades that followed, a massive number of incredibly old items were discovered, studied and cataloged. Those items included weapons, baskets, containers food storage containers, slings, and even "duck decoys" for hunting operations. Although archaeologists concluded that various tribes may have inhabited the caves over the years, certainly the most documented presence is that concerning the Paiute people, who flourished in not just Nevada, but also Arizona, California, Utah, Oregon, and Idaho. They continue to flourish. Not only that, they have a most intriguing legend – one of monstrous proportions.

According to the Paiute, in times long, long gone, they waged war on a mysterious race of giant humanoids known as the Si-Te-Cah. They were massive, violent, rampaging humanoids that fed voraciously on human flesh. Reportedly, the last of the Si-Te-Cah in Nevada were wiped out in the very heart of Lovelock Cave. They were forced into its depths by the Paiute, who filled the cave with bushes and then set them alight. The man-monsters reportedly died from the effects of fire and smoke. It was the end of a reign of terror that had plagued the Paiute for eons.

While there are rumors of at least *some* remains of the Si-Te-Cah being found in Lovelock Cave in the early 20th century, such a thing has not been fully confirmed. Granted, there are a lot of stories, but the skeletal remains of huge humanoids whose heights ranged from six-and-a-half-feet to *twelve*-feet? Well, that very much depends on who you ask. While there are no formally confirmed remains of such monstrous goliaths, stories certainly circulate to the effect that when the initial excavations began in 1912, the remains of a man who stood in excess of six-feet, and who was covered in red hair, were found – apparently in mummified, preserved states. So, the legend goes.

Of course, the reported physical appearance of the beasts – that they were humanoid, very tall, and covered in hair – has inevitably given rise to the possibility that, millennia ago, the Paiute waged war on a dangerous tribe of what they called Si-Te-Cah, but that we, today, would refer to as Bigfoot. A battle to the death, deep in the heart of Lovelock Cave? That just might well have been the case. It's no wonder, then, that the saga of Lovelock Cave intrigues and fascinates monster-hunters and cryptozoologists. It may also attract the attention of the Men in Black.

With the strange tale, and the even stranger history of Nevada's Lovelock Cave told, it's now time to return to David Weatherly who opened up about his very own, odd experience at Lovelock Cave. David said, as we awaited the call from *Coast to Coast AM*:

> This would have been May of 2016. [Researcher and writer] Dave Spinks and I were doing a series of investigations across Nevada and we had driven out to Virginia

City and a couple of other areas. While we were doing these investigations, we spoke to a Paiute elder at Pyramid Lake [Author's note: which is approximately forty miles northeast of Reno] who told us the whole legend of the Paiute version of Bigfoot, the Si-Te-Cah.

Ironically, when Dave and I were driving back across Nevada, heading east, I happened to pull off the interstate and realized when we pulled off the exit that we were in Lovelock. There was a little sign there with Lovelock on. The Paiute elder had not mentioned the name of the town. When I saw it, I was like, "Oh my gosh, Dave, this is Lovelock. This is where the cave is." I thought, we've got to find it. So, we drove into the downtown area. We went to the local library and asked for directions. It's almost twenty miles out of town and you literally turn off the pavement after about a mile, and you're on dirt road. The librarian said there was an archaeological sign there to look out for. But, it's in the middle of nowhere. She said, "Nobody ever goes out there." This was in the middle of the week; it was a Thursday. We were pretty sure there wasn't going to be anybody out there. So, we start this drive.

It was road-trip time – a road-trip that proceeded to get infinitely odd. And very quickly too. Back to David...

We get maybe a third of the way out there on these dirt roads; there's nothing, no houses, no buildings, no nothing. And, I'm cruising along, I was driving. And, I habitually check my rear-view mirror. All of a sudden there's a car behind us. It was really strange because there were no other turn-offs or anything. Dave was shocked, too. We were puzzling over where in the world did this car come from. It got stranger because this car maintained a regular distance between us. I just felt odd about it. It was a regular car; like a small Toyota. So, I hit the gas on the jeep, and I got up to probably 50, 60, 70 miles per hour on a dirt road. This car maintained the same distance behind us. He sped up and kept the same spacing. Which in itself was really odd. I'm kicking up a ton of dust, you know? So then I slowed way, way down. And again, he maintained the same distance.

Now, the road out there is a pretty straight stretch. But, it finally comes to a point where there's a big bend that hooks to the left. When I turned, the road started to slope up – up towards the cave. When I turned that bend and started to climb, I looked back and this car was gone. It had just disappeared. We get out to where the cave is, and there is a small parking lot and sign from the archaeological dig they did there many years ago. And, there's a little bathroom, but that's about it. So, we pulled into a parking space and both of us jump out.

We've got binoculars and we're looking for this freaking car. Well, we spotted it. Right before that bend, it had pulled off the road into the brush. It was sitting on the side of the road. Now, it's all the more puzzling. I'm thinking: what in the heck is this person doing? So, we think, we're not on a schedule or anything, so we decided we're going to stay in the parking lot and see what this guy does. It was probably twenty minutes to a half-hour before finally this car comes creeping up. The car pulls into the parking lot, There are only a handful of spaces in this lot. He parks it almost as far as he can from where my jeep is. And out jumps this young Asian guy. He has a hat on and he kind of jumps out of the car and he looks around with an odd gesture, and he says, "Aren't there supposed to be caves or something up here?"

Dave, who is closer to him, just points as it's pretty obvious from the sign. Dave says, "Right there." He pulls his hat down, rushes around to the trunk of his car and pops the trunk open. This was in the Nevada heat; this kid digs around the trunk and gets a coat, a flannel coat. The whole time he's trying to hide his face. He doesn't want us to see his face. We were trying to take pictures of him - very covertly. Then, he goes over to the billboard; there's this little artistic drawing and it says what life was like for the natives. It's a painting of a few native people, a handful of little comments about what they did, how they lived. He stood there for probably another twenty minutes just staring at this thing. Then he shouts over to us, "I guess I'll go up there." And he takes off."

That was hardly the end of the matter though.

Now, the trail does a big loop, if you want to take the long trail to get there. It's about a quarter of a mile. So, he takes off up this loop. We watch him and he gets around this rise and it kind of goes over some rocks and then he's gone – because the trail goes down the other side. Then it comes up and approaches the cave from behind. So we waited, again, for another half hour, because we wanted to do some filming for a project we were doing. Then we decided to hit the trail and we're filming as we walk.. David had the camera and I'm talking a bit about the caves and the legend. We get up to the rise and the trail hooks around to the left and we're still walking and all of a sudden this kid is behind us. It was the strangest damned thing.

> This kid had actually snuck off the trail and was hiding behind some rocks, waiting for us to go by. So, we just can't believe this. We stopped and he's glancing back at us. And every time we stopped, he stopped. He'll bend down, pick up a rock; look at a weed, or something like that. And, again, it's very forced. So, we do about two-thirds of the trail and there's an outcropping of rocks to the left. And to the right there's a lookout point. We climb this outcrop and we're sitting out and we just waited. And this kid, he hemmed and hawed, trying to delay and looking at things and finally, he goes passed us. He goes over to the lookout point and he kept glancing over at us.

Then, things got very surreal, as David makes abundantly clear:

> We're taking pictures on our cell-phones and he glances at us at one point and sees us with the phones. Then, he reaches into his pocket and pulls out a cell-phone. And this is what he's doing; looking at us as if he's trying to understand how to use it. He's stealing glances at us, like to make sure he's doing it right. Very weird. He stands there at this lookout point for a while, pretending to take pictures. Finally, he continues on. We let him go and we wait and we wait. Probably another twenty minutes; maybe half an hour. Then, we slowly make our way up to the rest of the trail, where the cave is. Figuring that by now surely this kid has gone, as the cave is small. As we get to the entrance of the cave, he pops out like some mad cuckoo. He rushes passed us and runs down the trail; he's in a sudden hurry to get out of there. We spent probably an hour and a half or two hours in the cave, because we wanted to check out a lot of different things related to the reports. And we did

some filming, too. Took a lot of photographs inside and out. We took our time.

Well, when we finally left the cave, and we took that corner, where you can see the parking lot; there's that kid sitting in his car. And as soon as we turned the corner, and started to trek down, he hits the gas and goes flying out of the parking lot. As if, again, he waited to see exactly how long we were going to be there and what we were going to do.

It's funny, some elements reminded me of the John Keel story of the Oriental Man in Black who stole an ink-pen [a story told in Keel's *The Mothman Prophecies*]. I should add, this car he was driving had no license plates at all.

We make our way down to the jeep and head back to town. And, when we get to about where the pavement starts, going into Lovelock, another vehicle comes toward us – pulls over very dramatically and suddenly onto the side of the road. Now, this is an all-black vehicle, with blacked-out windows. By now, we're not paranoid, but we are thinking there's something really odd going on here. So, I'm sitting at a stop-sign, watching this car do a very sudden U-turn to come and get behind us. I turned at the stop-sign, and the entrance to the freeway is right there, so I jumped on the interstate. This car was coming up behind us. Basically trying to catch up; he got caught up at the stop-sign as there was some other traffic. But, I left him back somewhere.

At the least, this is all weird synchronicities. The final anecdote is that from the time we were there at Lovelock Cave and had all these weird incidents, when me and Dave now speak on the phone, we get weird interference. Noises, clicking sounds. It doesn't happen when he speaks to other people; it doesn't happen when I speak to other people. But, it does if we bring something up about the cave."

It wasn't long before *Coast to Coast AM* called, and David immersed himself in the interview; while I pondered on all of those strange and off the wall MIB-themed experiences David had undergone. Two hours later, the show was over: David hit the road and I hit the sack. It had been an amazingly weird night.

24

"*THE MOTHMAN PROPHECIES* HAS BEEN INCREDIBLY INFLUENTIAL IN MY WORK"

August 2, 2017

"It kind of weirded me out"

Colin Schneider is someone who I run into now and again at Cryptozoology-themed conferences in the United States. At the age of just seventeen (as I wrote this), Colin has made great advances, in terms of researching and writing in the field of what is generally known as monster-hunting. He also has his very own radio-show: *The Crypto-Kid.* Colin happened to attend a conference I was speaking at in Ohio on the weekend of April 29, 2017. And, it was while we chatted that Colin briefly told me of a couple of weird experiences he had had, several of which made me think that certain, supernatural things had got their grips into Colin, and in much the same way that they had gotten into John Keel. And into me, too.

We're talking about curious phone calls and the appearance of the highly disturbing phenomenon of the so-called "Phantom Social Workers" (alternately known as "Bogus Social Workers"). They visit the homes of those who have had supernatural experiences, but do so under the guise of working for government agencies. I wrote extensively about the PSW and the BSW in my 2016 book, *Women in Black.* As for Keel, back in the 1960s, he found himself trailing the very similar "Phantom Census

Takers." They were pale-faced women who claimed to be doing research on the part of the U.S. Government. They were, however, absolutely nothing of the sort. Whatever they were, and from wherever they came, the PCTs were near-identical equivalents of the BSWs and the PSWs.

It was in August 2017 that I was finally able to get around to interviewing Colin, whose intriguing story follows:

> When I was about eight, there was a TV show, *Monster Quest,* on the History Channel. At that age, everything was about monsters. It just caught my interest because at the time I was very into animals. I used to go out and collect different reptiles and amphibians and put them into this big fish-tank that I had. So, the idea that these monsters might be living animals, just waiting to be discovered, really got me excited and I started to read a lot about the subject of cryptozoology. On pretty much every episode, there was this guy named Loren Coleman. He always seemed to know everything on every topic you could imagine in Cryptozoology. So, through watching him on *Monster Quest,* I started getting a lot of his books, and I really started looking up to him.
>
> I was thirteen when I went to his International Cryptozoology Museum and that just got me incredibly electrified about the subject. And I got to meet Loren Coleman and we chatted for a while. I remember walking out with a stack of books, thinking that this is what I want to do with the rest of my life. And thinking that I want to be a monster-hunter. I want to find these things. Since then, I've gone to conferences, researching and writing, and now doing my show.
>
> When I first started, it was all about Bigfoot. It has always

> been a favorite of mine. But, as I slowly started to mature in my research, and started to learn about the nitty-gritty of the subject, the Mothman really caught my eye. It was something that was so weird, and so out there, that it really fascinated me. John Keel's *The Mothman Prophecies* has been incredibly influential in my work, and what kind of researcher I have turned out to be. I read the book a couple of times and each time I read it I got something new out of it. It kind of shifted the way I looked at things. For a long time I was very into the idea that all of these things are flesh and blood animals or they don't exist. But, Keel's book changed all that.

It was when Colin got into the strange world of John Keel that even stranger things began to occur, as Colin notes:

> The second time I read *The Mothman Prophecies,* I was just about done with the book and, as I put it down, I was sitting there and I got this phone call. I answered it and I just heard these strange sounds on it. It said "Private number" and I normally don't answer those, but I did this time. It was just really weird static-type sounds. No-one said anything; just static. And, it kind of flipped me out, because in Keel's book that's one of the big things John Keel reported having occur: Men in Black, static, and weird phone calls – and it kind of weirded me out. This was June 2014.

Things proceeded to get even weirder for Colin, as he revealed to me:

> There was something else in the summer of 2014, which happened when I was really getting into this field. This was also the summer I went to my very first conference. So, I was getting really excited about this. And, I was over at a friend's house and at the time I was staying with my aunt, on my mom's side. During the summers, I stayed with my mom quite a bit because of the way situations were – my parents are divorced. I remember coming back to my aunt's house and hearing from my aunt that just previously two social workers came to the house and were talking to the kids.
>
> That's nothing unusual; my aunt had a lot of problems with her kids and the ex. So, there were social workers who came over quite a few times. So, it wasn't anything unusual in its own right. But, what was weird, was one time when they sent out different social workers; normally they send out the same people, because they kids feel more comfortable with them. They were two women and they asked to speak with the kids, and sat them down in the living room, chatting with them. They said everything seemed to be okay, and they would have them come back to the office for a follow-up. They said they would call in the week. One, they never called. Two, after they didn't call my aunt, she called the office and they said they never sent anyone out. What really made me think of this was when I read your *Women in Black* book and read about the Bogus Social Workers in the U.K. – I assumed this was just a U.K. thing. So, it didn't click until I read your book.

None of this, fortunately, adversely affected Colin's enthusiasm for the world of the unknown, as his following words to me make clear:

My biggest plans now are to continue going to conferences, giving talks, and to finish my book. Long-term goals: I want to get a degree in zoology or anthropology and if I do go with zoology, I want to work with a zoo. And to continue doing my radio show, as I love doing that.

25

"THE DARK DRESSED MAN"

August 28, 2017

"The men exuded an uncanny energy which I immediately sensed"

One of those who had a table at the 2016 Mothman Festival was Susan Sheppard. I first became acquainted with her when she appeared on a 2010 DVD documentary titled *Eyes of the Mothman.* It's an excellent, moody and mysterious study of the Mothman phenomenon – with the MIB getting significant coverage, too. Susan and I had a brief chat about Mothman and the MIB and she very generously offered to share with me her personal MIB-based experiences – as well as additional material on Mothman. In fact, Susan did far more than that. She prepared for me an extensive, historical account of her very own knowledge of the MIB-Mothman phenomena which hit Point Pleasant, West Virginia in the 1966-1967 period.

Just like I did on August 28, 2017, you'll now quickly come to see that Susan has prepared a powerful and unique report on her very own memories, experiences and much more connected to the mystery of the Men in Black – as well as on a wealth of related events which occurred in and around Point Pleasant, West Virginia, when the MIB and Mothman were both ever-present on those dark and disturbing nights. In fact, Susan – as you will soon learn – may well have had her very own encounters with a somewhat MIB-like character that haunted the area back in the Sixties and who went by the somewhat odd and Dickensian name of Indrid Cold.

Weirdly, on the same day that Susan emailed me the document, something very odd happened: I had countless sightings of what I can only describe as small, shadowy things scuttling around the rooms of my apartment, and at the very edge of my peripheral vision. It was as if ethereal, hard to define, rat-like animals were swarming around at incredible speeds – taunting me by not making themselves entirely visible, but also determined to be sure they could be seen to at least some small degree. I also had weird and uncanny feelings of being watched – by who or by what I could not say. But, put it this way, I certainly didn't get a good vibe from what briefly went down.

On the night of August 28, I cracked open a cold can of *Steel Reserve*, stretched out on the couch, turned the volume down on the TV, and proceeded to read. And, now, it's your turn to do exactly that, too.

Enjoy...

WHAT I KNOW OF THE MEN IN BLACK, THE MOTHMAN & INDRID COLD

– BY SUSAN SHEPPARD

Merle Partridge and the "Red Eyes in the Barn"

I may be one of the few people who remembers the West Virginia Mothman and related events before the "Mothman" even had a name. It was one chilly November day in 1966 that my sister came home from school to tell a peculiar story that happened the night before to Merle Partridge's family. I was small at the time but I remember it this way: I sat at the kitchen table eating what remained of Halloween candy from the bottom of a grocery store bag. There was plastic over the kitchen windows to keep the cold out. The hill outside was covered with brown and yellow grass that seemed to ripple. It always seemed like there were presences in our woods. My sister burst through the front door and then told a story about "Paula Partridge's Dad" seeing red eyes in the barn the night before and that their German-Shepherd dog Bandit was now missing.

Strange red eyes staring out of the doorway of an old barn wasn't something you heard about often in central West Virginia, so my childish mind was awakened and I held on to the story probably longer than anyone else in my family cared to. I can tell you that your ordinary West Virginia citizen didn't believe the tale at first. Until these strange manifestations that became Mothman, Indrid Cold and the Men in Black began to touch their lives in both big and small ways.

My family lived on Shannon's Knob in the small town of West Union, in Doddridge County, West Virginia, while the Partridges lived about eight miles away from us in a country community called Center Point, basically across a ridge or two. My grandmother had been the post mistress of Center Point, where she grew up in a log cabin. There was even a Banshee story associated with the community.

Although the dates given vary, it is usually accepted that the day of the Merle Partridge sighting or the "red eyes in the barn," happened on November 15, 1966, was the same night as the Scarberry encounter with the Mothman an hour and a half later near the TNT plant in Point Pleasant, which was about 100 miles away from Center Point.

Shortly before 10 o'clock in the evening, Merle Partridge was watching late night television with his son Roger, who was about 11 years old at the time. They first noticed their German shepherd dog barking outside, it seemed at a distance from the house, in an atypical kind of way that made them think something was terribly wrong. At that same time, the television set started "screaming" as the picture blanked out and the TV made a loud grinding sound, which Mr. Partridge later told author John Keel, "sounded like a generator winding up."

Partridge's account, which he told me (Susan Sheppard) in an interview in 2006, differed slightly than what John Keel eventually included in his *The Mothman Prophecies.* First, Keel must have misheard Merle Partridge's name and wrote it down as "Newell Partridge." Mr. Partridge was always open about the story and never asked to be given an alias. Merle Partridge told me, "I don't know why John Keel wrote my name down as Newell." It's possible the New York City author couldn't understand Partridge's West Virginia accent.

After listening to Bandit bark for a little while longer, Merle Partridge decided it might be a good idea to investigate the noises. It could have been a prowler on his property or even a black bear. Center Point was miles from any town. Mr. Partridge grabbed a flashlight, took Roger in tow and the two headed outdoors to find out what all of the ruckus was about. Partridge didn't see any red eyes from a distance, as some accounts claim. Instead, he saw Bandit standing at the entrance to the empty barn, which Partridge described as "a football field away." The dog barked frantically as he continued to stare into the barn.

As Merle Partridge and Roger got closer to the barn, Partridge said he felt the hair raise up on his arms. When he looked into the barn, he saw what he described as what looked to be, "Red, rotating electrical lights." Partridge explained to me, "Red lights, red eyes, whatever you want to call them."

On that chilly November night Partridge noticed what he thought was a dark shape lumbering up from the floor of the barn. The fur stood up on Bandit's back, the dog snarled angrily and he shot into the barn toward whatever the dark form was.

Merle Partridge and Roger ran back to the house. As they sat down in front of the television set, Bandit stopped barking and the picture came back on the screen of the TV. Calm was restored. Still, Mr. Partridge later mentioned that he slept with his rifle beside of him that night. The next morning, Bandit did not come to the back door for his breakfast as he usually did. Merle Partridge remembered that the last time he'd seen Bandit, the dog was running into the barn seemingly in pursuit of the red eyes, or "lights" and the dark form.

Merle and his children Mary, Roger and Gary headed for the barn and what they found inside was chilling and mysterious. They discovered Bandit's paw prints in the dirt floor of the barn, but the paw prints only went in a circle and didn't lead away. It was as if that dog had been picked up and carried away by something much stronger and larger. Mary Partridge also commented there were other tracks, but they didn't belong to the dog. She said they looked like giant turkey tracks but ones like she'd never seen before.

At this point, Merle Partridge didn't realize that the night before, less than an hour after his encounter with the "red eyes in the barn," Roger and Linda Scarberry, of Point Pleasant, along with Steve and Mary Mallette, would also meet up with these "red eyes" at the TNT plant on the outskirts of town, and then more times along Route 2, except now they were able to see a figure attached to the red eyes or "lights" in detail. It was a tall humanoid creature of over six feet tall, with crimson eyes that seemed to be set in its shoulders, and a wingspan of almost 10 feet. (Later witnesses would describe the creature as varying shades of grey, brown or tan and sometimes flesh-toned.)

In later interviews, Linda Scarberry, like Merle Partridge, described the eyes of the creature as looking like "red lights" and not really the bicycle reflectors that Keel compares them to in his book. In fact, at one point when the Scarberry's thought they had outdistanced the Mothman after he had been flying over their car, Linda looked into a field and thought she saw red lights on a billboard. When the billboard was caught in the beams of the Scarberry's car, Linda claimed the Mothman was perched on the edge of the billboard. As they passed, the Mothman once again took flight and continued in pursuit of their car.

As they came upon the city limits of Point Pleasant, Linda Scarberry looked and saw the dead body of a large dog beside the road. As the young couple drove, the Mothman soared overhead but only a few feet above the car. The wings of the creature were so huge that as they flapped they hit the side doors of Roger's car. Once they arrived in town, they drove directly to the Point Pleasant Police Station and made their famous report. Sources say it was Roger Scarberry who made a sketch. When the police went out to check the car, they found large scratches on each side door.

When finally the frightened couple returned to their house, Linda Scarberry claimed the Mothman had followed them and peered into the windows the rest of the night. In one interview Linda remarked, "Even to this day, I will not look out my windows after dark."

Within a day or two, the Mothman story (the creature did not yet have a name and was referred to as a "bird" or "Birdman) made its way through news syndicates, appearing in newspapers but mostly in the mid-Atlantic region. These caught the eye of Merle Partridge in Doddridge County because the account seemed eerily similar to what he had experienced on the same night, but what he really noticed was the mention of the dead dog beside the city limits sign of Point Pleasant. The Partridge's German shepherd dog Bandit did not return, nor would he ever. Merle called the News Syndicate who put him in touch with Mary Hyre, where the Doddridge Countian gave his report. Mary Hyre wrote up Partridges account and kept it. Later, John Keel would follow up on the story and include it in his book *The Mothman Prophecies.*

Who's Afraid of Indrid Cold?

But the strangeness that enveloped West Virginia and Ohio Valley did not begin with the Scarberry encounter along Route 2 north of Point Pleasant. There had been another puzzling event that occurred in Parkersburg 12 days before. This happened on November 2, 1966, one mile south of the city limits of Parkersburg. It involved a sewing machine salesman whose life was about to be disrupted in such a way that he would never entirely recover.

His name was Woodrow Derenberger, but everyone called him "Woody." It was shortly after 6 p.m. in the evening, when Woody Derenberger was driving home from his job as a sewing machine salesman at J.C. Penny's in Marietta, Ohio to his farmhouse in Mineral Wells, West Virginia. The ride home was overcast and dreary. It was misting a light rain.

As Derenberger came up on the Intersection of I-77 and Route 47, he thought that a tractor trailer truck was tailgating him without its lights on, which was unnerving, so he swerved to the side of the road and much to his surprise, the truck appeared to take flight and seemed to roll across his panel truck. To his astonishment, what Derenberger thought was a truck was a charcoal colored UFO without any lights on. It touched down and then hovered about 10 inches above the berm of the road. Much to Derenberger's surprise a hatch opened and a man stepped out looking like "any ordinary man you would see on the street - there was nothing unusual about his appearance."

Except the man was dressed in dark clothing and had a "beaming smile." As the man proceeded to walk toward Derenberger's panel truck, the "craft" jetted up to about 40 feet in the air where it floated above the highway. What happened next was unsettling, because as the darkly-dressed man came up toward the vehicle Woody Derenberger heard the words, "Do not be afraid, I mean you no harm, I only want to ask you a few questions." Derenberger did become afraid because as the man spoke to Woodrow, his lips did not move. The man then moved to the opposite side of the truck and told Derenberger to roll down his window so they could talk better, which he did. Next what formed in Derenberger's mind were the words, "Now you can speak, or you can think... it makes no difference, I can understand you either way," ... this is what the dark man said.

Later, when Derenberger was questioned on local live television, he was scrutinized over what seemed a contradiction because if the dark man communicated through a type of mental telepathy, why would Derenberger need to roll down his window to talk? Wouldn't it be easier just to talk mentally?

Woodrow Derenberger explained it was because Indrid Cold wanted to look directly at him as they spoke and he felt that, really, Cold wasn't so interested in what was said but more interested in keeping up a communication with him. To Derenberger, that seemed the entire point of it all. Derenberger also noted that when Cold stared into his eyes, it was as if he knew everything about Woodrow Derenberger, and also, if he could only let go of his fear and do the same, he felt he would also know and understand all about Cold. In any event, Cold spoke through the passenger side window the entire time.

The physical description of Cold was commonplace. Derenberger described him as about 35 years of age, having a trim build, was about six feet tall, 185 pounds with dark eyes and dark hair slicked straight back. Cold wore a long dark coat and beneath the coat, Woodrow Derenberger was able to glimpse the fabric of his "uniform" that glistened beneath the coat. He also described Cold as having a "tanned complexion." Throughout the conversation, Cold kept a frozen smile and curiously hid his hands beneath his armpits most of the time.

Cold did, however, point at the city lights above the distant hills of Parkersburg and asked Mr. Derenberger, "What do you call that over there?" Derenberger said, "Why, that's Parkersburg and we call that a city." Cold responded, "Where I come from we call it a gathering." Cold later added the curious statement that "I come from a place less powerful than yours." As the men talked cars passed under the craft, it drifted above the road. The occupants were seemingly unaware of the spaceship being there. After all, there were no lights that could be seen. Cold then asked about Parkersburg, "Do people live there or do they work there?"

Woody Derenberger answered, "Why, yes, people live and work there." Cold interjected, "Do you work, Mr. Derenberger?" (Woodrow told Cold his name as the conversation began.)

Derenberger answered, "I am a salesman. That's what I do. Do you have a job?" Cold answered, "Yes. I am a searcher."

After that the conversation became mundane. Cold seemed to notice Woodrow Derenberger was scared and commented on it. Mr. Derenberger claimed Cold asked him, "Why are you so frightened? Do not be afraid. We mean you no harm. You will see that we eat and bleed the same as you do," and then added an emotive note, "We only wish you happiness" which Cold said to the frightened man more than once.

While Mr. Derenberger was being interviewed on live television on WTAP-TV, he attributed this puzzling statement to Indrid Cold, "At the proper time, the authorities will be notified about our meeting and this will be confirmed." The entire conversation took between five and ten minutes and then Indrid Cold looked inside Woody's car with his ever-present smile, and said, "Mr. Derenberger, I thank you for talking to me. We will see you again."

He ended the conversation with "we will see you again" and as soon as he did, the spaceship immediately came back down, floated about 10 inches off the road. A hatch opened and a human arm extended pulling Cold up into the craft. The ship then jetted up into the air about seventy-five feet, made a fluttering noise and then shot away at a very high rate of speed. For a few moments, Woodrow Derenberger sat stunned. Finally, he started up his car and drove to his farmhouse in Mineral Wells where his wife met him at the door. By now, it was shortly before 7:00 o'clock.

Mrs. Derenberger met her husband at the door. She later said that Woodrow "could not have been any whiter if he had been lying in a coffin." The stories vary but from Mr. Derenberger's account his wife is the one who called the West Virginia State police, or at least she dialed the phone. Woodrow Derenberger gave them a brief report of what he claimed to have happened. It is interesting to note that in the initial report, Derenberger called the alien "Cold," but did not mention "Indrid" until later.

The next day Derenberger attempted to go back to work but was sidetracked when he agreed to a live television interview about his experience on the previous night with a UFO with WTAP-TV the NBC affiliate in Parkersburg, housed in a small building not much bigger than a garage. The interview took place shortly before noon where Woodrow Derenberger was grilled by veteran reporter Glenn Wilson and city Police Chief Ed Plum, as well as other local law enforcement, including the head of the Wood County Airport. Representatives from Wright Patterson were in route to interview Derenberger but whether that came about is not known.

The interview went on for about two and a half hours. The live part of the broadcast was under an hour long and then the television cameras were turned off, and the interview continued off the air for another hour or so. During that time, Derenberger drew a picture of the spacecraft that he described in his thick West Virginia accent as a charcoal grey, with no lights and looking like an "old-fashioned chimney lamp." (You may want to google this because I am not clear what he intended. I found some interesting images when I did...Parts of the lamps can look like UFOs.)

Probably one of the most curious statements Woodrow Derenberger made about his meeting with Cold was, "And then Cold said to me, we will see you again..." then his voice trails off. Police Chief Ed Plum asked, "Do you really believe you will see him again?" Derenberger then answered, "I think so...I believe I will...I don't know... because that's what I am afraid of."

After that interview, Derenberger's life transformed drastically and not for the better. He changed jobs, developed marital problems, clung to his church for a while, and then came the strange visits from men dressed in black clothing whom Derenberger suspected to be some kind of hidden government group of spies or maybe even the Mafia. He wasn't sure, they just spooked him. They would arrive his house, ask Derenberger simple questions, (some had to do with his UFO experience) and then the Men in Black acted in a threatening manner.

But nothing was as incredible as the return of Indrid Cold. At least, this is what Woodrow Derenberger claimed. He said that

Cold visited him many times at his farmhouse in Mineral Wells. At one point, Derenberger came up missing for almost six months and said he was "with the aliens." The local population finally became skeptical. The sewing machine salesman's tale grew more and more far-fetched. Derenberger even claimed to have been impregnated by the aliens. In 1967, Woodrow Derenberger stated to have visited Indrid Cold's home planet of Lanulos where its residents walked around wearing no clothing. He said the aliens lived in a galaxy called Ganymede where everything was peaceful and there was no war. People began to snicker.

Still, there were odd flashing lights in the sky almost nightly and the curiosity seekers stalked not only Derenberger's modest farmhouse, but an area called Bogle Ridge, not far from Mineral Wells where the aliens were claimed to land. The ridicule became too much. Derenberger, with his family, moved from the area and stayed away for decades. He returned to Wood County in the 1980s and died in 1990. Woodrow Derenberger was finally laid to rest at Mount Zion Cemetery in Mineral Wells, West Virginia.

John Keel was not a believer in Woodrow Derenberger's UFO story, so it's mysterious why he would make it such a big part of *The Mothman Prophecies* book. In *The Mothman Prophecies* movie the character Gordon Smallwood is based upon Woodrow Derenberger, but the Wood County man most often appeared in a suit and not overalls. A few elements to his story make it believable that, initially, something of an extraordinary nature happened to him. First of all, his account predates the Mothman sightings by 12 days. Derenberger would have to have been a prophet to know what was about to happen next, making his story even more extraordinary. His family explains that they believe something of an otherworldly nature initially happened but he added to the tale to sell books when he self-published a book called *Visitors from Lanulos* in 1971.

There are a few other accounts that add some believability to key aspects of Woodrow Derenberger's fantastic story. An elderly man driving south of Parkersburg on I-77 reported seeing a man by the side of the road (one that meet Indrid Cold's description) who tried to flag him down. The gentleman slowed down but when the darkly clad man headed for the passenger door of his car, the senior citizen became frightened and drove away. "There was something off about that character," the old man later told Glenn Wilson of WTAP-TV.

I also ran into something curious when I was researching stories for my ghost tour back in 1996. I found the news article about Woodrow Derenberger's UFO tale in the *Parkersburg News & Sentinel* dated November 4th, 1966, where the story was on a front section of the newspaper. The account read: "Local Man Stopped by UFO." In the same section of the newspaper, right beside of it, was another smaller article about a complete power outage that happened in South Parkersburg at precisely the same time Derenberger claimed he was interrupted by Indrid Cold's spaceship. South Parkersburg borders the community of Mineral Wells. An energy disturbance in Mineral Wells would likely also affect South Parkersburg.

My Grandfather and his Huge Ball of Light

My Grandfather, W. P. Chapman of Cairo, West Virginia in Ritchie County, worked as a signal man for the B & O Railroad. In the fall of 1966, he was checking signal lights that weren't working properly late one night outside of Cairo .He was driving Park Road near Cairo when suddenly what looked to be a fiery ball of light came rapidly up behind his truck .As the ball of light rolled across top of his truck, the power went out and the truck rolled down into a dark field. My Grandfather's vehicle was completely drained of its power. The ball of light flew forward then disappeared into the blackness of the night sky. As the ball was out of sight, the power in his truck came back on. Unnerved, my Grandfather drove on to check the signal lights which now appeared to work even though before the lights had been reported broken.

In late February of 1967, my Grandfather was sitting at his desk in front of a large window in Cairo. As my grandmother passed behind his chair, she noticed three bright lights in the evening sky: red, white and green. They blinked on and off at different times, then disappeared. A few weeks later, my Grandfather was again checking faulty signal lights. Unfortunately, he suffered a heart attack on the train tracks and was unable to get out of the path of an oncoming train. My Grandfather was killed. My West Virginia Grandmother later described the lights in the sky as "tokens" or warnings of his death, but UFOs that were proliferating in the nighttime skies were much more likely than Appalachian folklore.

Men in Black on Shannon's Knob, Doddridge County, West Virginia

In late winter of 1967 a church bus was driving over Route 50; it then went directly through the small town of West Union, where I was raised. The occupants of the bus noticed othat a UFO was hovering over my family's home on nearby Shannon's Knob. Shannon's Knob was the hillside I grew up on and the highest point in the town. A woman who was a passenger on the bus called my parents to tell them about the spacecraft suspended over our home but my parents were disbelieving.

The following spring, my friend Regina Ball and I were playing on the hillside above my family's house. At the peak of Shannon's

Knob was the entire power source for the town of West Union. We walked up a hill near a clearing to play "Indians" like we always did. As Jeanie and I played, we suddenly heard men's voices. Turning, we saw two men dressed in black near the rise above us. We were frolicking near the bushes but when we noticed the Men in Black we became scared and hid. We watched silently as the two men measured a spot on the hillside.

The men exuded an uncanny energy that I sensed immediately. They also looked different from each other. One man appeared to be of European descent while the other one had an Asian appearance. The one with the Asian appearance had what looked to be dyed blond hair that was cut very short. Neither was wearing a hat, nor did they wear a suit and tie. They simply had on a black shirt and black pants. The men seem interested only in the landscape and spoke to each other in low voices. Jeanie and I spied on the two men until they left. There was no vehicle anywhere close. Where they came from, we did not know. Jeanie and I went home and forgot all about them. In years to come, we would both suffer from migraine headaches and just overall poor health. We never linked the Men in Black up on the hill with our headaches. I later learned Derenberger suffered from severe headaches, as did his daughter.

A few days after some boys were playing on the same hillside. The boys noticed a circular impression in the grass that was about 15-20 feet wide in a treeless area on Shannon's Knob. They went home and reported it to their parents. My brother came home from school relaying the story from his friends. In my child's mind, I made the connection. It was almost like a secret that only I was privy to. To my knowledge, no one ever followed up on the circular impression on Shannon's Knob nor the UFO sighting over our house.

In the winter and spring of 1967, when my Grandfather was

hit and killed by a train and we witnessed the Men in Black up on Shannon's Knob, we had other bizarre happenings. Our home became an epicenter for poltergeist activity. Ceramic birds flew off our living room mantle. Pictures fell off the walls. A heavy iron was tossed from my bedroom into the bedroom of my parents and slid under the bed where my Dad was sleeping.

But nothing was quite as strange as the footsteps I heard walking on the roof at night. This was the winter of 1967 and for the most part, appearances of these footsteps went on for as long as up until 1970. On certain nights, around 2:00 a.m., a loud bang would punctuate the stillness of my bedroom. It originated from the roof above me. It sounded like someone had jumped out of a helicopter onto our roof with a boom. There would be a pause of a few moments, and then, whoever it was, began to walk on the roof. The roof would creak under the weight. If I would scream for my parents, which I did often, the footsteps would pause until it grew quiet again and they would start right back up. At first my parents did not believe me. However, one morning I heard them talking amongst themselves. My parents had heard the footsteps on the roof as well.

I began to sleepwalk. Sometimes I would wake up to hear the radio playing what sounded like a Catholic Mass but there were no active Catholic churches in my town. I reached over to turn off the radio and found that the radio wasn't on. Then I would lie there and listen to a "broadcast" of an "angel choir" with no source for the music until the cords faded as the sun finally came up.

One morning, I woke up in a bedroom my family never used, with a sheet pulled over my head like I was a dead person on a gurney ready to be taken to the morgue. As I stirred from sleep, I felt my skin had grown cold against the clammy air and realized every stitch of my clothing had been removed. I had no idea how I got there nor why my clothes had been taken off me. Embarrassed, I jumped from the bare bed, put my clothes on and never told a soul. My parents were sleeping normally in theirs. I was seven years old.

Perhaps the strangest tale of all was one about our house; it happened more than a decade before we moved in. A young woman was babysitting at what would later be our home. She thought she heard something attempting to climb up the wall of the house. Too frightened to go check for herself, the girl called the West Union City Police. When cops came to check the house, they were shocked when their flashlights shone up the side of the wall. There were muddy footprints traveling up the side of the house. Whether they were human or something else, that part I don't know. From then on, our house became known as one that was "haunted" in town. But by what?

It Didn't End There for Merle Partridge

Merle Partridge and Woodrow Derenberger had a number of things in common. Both were every-day-working-family men with unexceptional lives until the month of November 1966. Derenberger was a sewing machine salesman and Merle Partridge worked as a truck driver. After going through such mysterious occurrences, each man's life turned the page to a new chapter of high strangeness.

Merle Partridge never sought the limelight in the same way Woody Derenberger did. He made his report to the news syndicate and was later interviewed by *Mothman Prophecies* author John Keel and that was it. Partridge was done with the story. Keel took some liberties with Partridge's account. First, he wrote "Merle" as "Newell" in the book, which Partridge never fully understood. Partridge also thought the "red eyes" actually looked more like red lights that rotated and circled, not like bicycle reflectors that Keel put in the book. Lastly, Partridge lived at Center Point and not nearby Salem, although Salem may have been Partridge's mailing address.

In 2006, I interviewed Merle Partridge. He didn't have much to add to the initial story of Bandit and the red orbs in the barn other than what had been published, and as it was with most Mothman accounts, what happened to Partridge was memorable but very brief. Except for the stories that came later.

Merle claimed that in the spring following his experience in November, he was lying out on the deck of his house relaxing when suddenly the sky darkened. He looked up to see a dark grey spacecraft moving soundlessly above him. He saw no flashing lights but said the craft was so enormous that it cast a shadow over him and the house. Partridge also claimed the UFO made no noise whatsoever and that it was gone within minutes.

It wasn't but a few weeks later that Partridge had a knock on his door around 9:00 p.m. He opened the door to find a middle-aged man who was clearly upset. The man explained he had run his jeep into a ditch a few minutes before, but that wasn't what he was worried about. He said he could not find his six-year-old son who had been sitting in the car seat beside of him. The man said as the two drove down the dark country road, something dark flew over and blocked out the view through his windshield. The next thing the man knew his Jeep was sitting in a ditch and his son was missing from the passenger seat.

First thing Merle Partridge and the man did was go out searching for the six-year old with flashlights, but the boy was nowhere to be found. Partridge then dialed the Salem police department and they made it to the farmhouse within a half hour. With two policemen, they once again embarked to search the country road and grounds where the boy came up missing.

Much to their surprise, the young lad came down the road in the opposite direction of where he and his father were traveling. It was as if the boy was sleep walking. When the father shook him and asked him where he had gone, the boy had no recollection of where he had been for the almost an hour- and-an-half. The night ended in both a relief and a mystery.

About thirty-five years later, a man who looked to be about forty years old knocked on the door. By this time, Mr. Partridge and his wife were living in New Martinsville, which is 100 miles upriver from Point Pleasant. The man introduced himself by a name that was not familiar and then said to Merle Partridge, "I know you don't recognize me but I was the boy that got lost after our Jeep went off the road that night thirty-five years ago and you helped find me. To this day I have no memory whatsoever of that hour I was lost."

The last story Merle Partridge told me did not disappoint in its peculiarity. Most of his life was spent as a truck driver. Mr. Partridge was on a particularly long drive, it was a foggy morning and he made it almost home when he was overcome by a terrible fatigue. He decided to pull over and get a few winks in the cab of his truck. After sleeping for a few hours, Mr. Partridge finally awoke to a sticky feeling on his hands and face. As he opened his eyes, he was shocked to find that his body was covered in cobwebs.

That ended the stories Merle Partridge told me in 2006. He was featured in the movie *Eyes of the Mothman* along with other Mothman witnesses but passed away before ever seeing himself on the silver screen. Merle Partridge was a kind and intelligent man who never sought out attention or special treatment. I believed all that he said.

The Derenberger Tapes and Enter Billi

I did monthly live horoscopes or an astrology segment on WTAP-TV in Parkersburg for a number of years, and knew Glenn Wilson, the man who had interviewed Woodrow Derenberger in November of 1966. When Glenn Wilson retired around 2001, he had something he thought I might like to have – they were the original reel-to-reel tapes of his live interview with Derenberger on November 3rd, 1966. These were audio tapes of more than two hours long and on them was written "UFO TAPES November 1966." Wilson almost threw them away, he told me, because he felt Derenberger had given Parkersburg a "black-eye" having made the city laughing stock over his alien visits and extraterrestrial pregnancy claims. Wilson said there was also video of the live interview and a drawing Derenberger had produced of Indrid Cold's spaceship but both came up missing. Wilson assumed the cleaning lady threw them out with the garbage.

The infamous UFO interview had not been listened to for about thirty-five years and hearing them for the first time was quite remarkable. Musician and Emmy winner David Traugh, who owned a recording studio in Parkersburg, transferred the rare interview to a cassette tape in the summer of 2001. Woodrow Derenberger had come back to life to tell his story all over again. One could hear him rap his knuckles on the table for emphasis during the interview and listen as he faltered a bit. Yet Mr. Derenberger was consistent in everything he said.

Later, I burned the interviews on to a CD and presented them to author John Keel in 2003 at the only Mothman Festival (held each year in Point Pleasant) he attended. He commented that he didn't even know the interviews existed and seemed skeptical that they were real. I assured him they were genuine and I hoped he would enjoy them. The last thing John Keel said to me was,

"I hope you make some money off of these." (Not really, I've given away more CDs than I've sold...in any event, that wasn't the purpose anyway.)

However, before all of this, in 2001, I was doing a book signing for my astrology book that was published by Kensington Publishing. A young man in a long dark coat appeared at the bookstore. He was dressed in dark clothing, was about six feet tall with dark hair combed straight back. He had a medium build and dark eyes. The man appeared to be in his mid-thirties and introduced himself as "Billi"... (He had me sign a book.)

Billi was very good-looking, in fact, he resembled the actor Richard Gere and I thought he had a Slavic or part Native American appearance. But what struck me instantly was Billi didn't appear as that intelligent. In fact, he acted rather dumb.

Billi picked up my book and asked me, "Where did you find this book?"

I answered, "Well, I wrote it and this is my book signing."

Billi turned the book over. "Does it have to do with stars?"

I said, "Yes, your sign, like when you were born. The month and date tells what your sign is."

He looked puzzled and commented, "I was born in November of 1966."

I then said, "That would make you probably a Scorpio, intense in nature, interested in subjects others may feel a bit off-putting."

Billi looked thoughtful. "I assist in brain surgeries. I have witnessed many. I like the way the brain looks when the skull is open."

"Oh, you're a doctor?" I asked. (My inner voice is now saying "How can this man be a doctor?")

Billi answered, "No. I am a helper."

"You mean a nurse?" I asked.

"Yes, I am a nurse, I guess." Billi giggled. (Inner voice "There is no way this guy made it through nursing school." Also, brain surgeries were not performed in the two local hospitals. Brain surgery patients were usually sent to Morgantown, Pittsburgh or Columbus.)

"Oh," I responded (Inner voice asks, "Why is this man lying to me?").

"I think I am going to buy this. Will you write your name on it for me?" Billi clasped the book to his chest and then took it to the cash register. He pulled some bills out of his pocket, paid for the book and brought it back to me.

"Put your name there," Billi commanded as he sat the book down in front of me.

I said, "Okay…but yours, too, what's your name?"

Billi said, "My name is Billi."

"As in B-I-L-L-Y?" I spelled the name out.

"No," said Billi, "there is an 'I' is at the end of my name."

I signed the book 'To Billi," and asked, "Is that the right way?"

Billie said, "Yes. That's correct."

"Well, there you go and thanks so much." I handed Billi his book.

"Thank you for talking. I have to go now…to go find my brother…he went somewhere. He may be lost, I think." Billi glanced over his shoulder, picked up the astrology book and vanished through the doorway of the bookstore.

When writing on speculative subjects, including astrology, you meet people from all walks of life and an odd person appearing can be normal in the strange, speculative world, so I didn't give Billi much thought other than to remember the peculiar encounter.

It was now August of 2001, so it was time to begin preparation for my seasonal ghost tour and that meant putting a new message on my answering machine. One night I walked in to find my answering machine's red light blinking (those were the days) and played back the message that turned out to be a hissing voice saying the word "Hi!" But the word was drawn out and the raspy greeting sounded more along the lines of: "Hiiiiiiiiii-eeeeeeeee-yaaaaaaaah!" It was really a long hiss.

I'm used to prank calls so I deleted the recorded message and carried on. But there was something about the voice that didn't seem exactly human; however, some people can do bizarre acrobatics with their voices and as had been my meeting with Billi, it was more of a "Welcome to my strange world" sentiment that I brushed off and partly forgot.

A Ghost Tour Season of Shock then Awe

In the following weeks a catastrophe would hit the United States on September 11, 2001, when the World Trade Center in NYC, Shanksville, PA and the Pentagon were attacked and 2,996 innocent individuals lost their lives. Like most Americans, I tried to move past my own shock and fulfill my obligations, and that would be preparing for another ghost tour coming up in little over a month. I thought maybe one new twist on the ghost tour might by playing some of the Derenberger / Indrid Cold interviews for the crowd. It would be not only an escape for everyone but also entertaining. In late 2001, aliens, as well as ghosts, were now pretty low on the list of scares.

The 2001 fall season grew cold rather quickly and tour goers were soon wearing winter coats in October. One night, I looked over as I told the Indrid Cold story and was about to play the Derenberger tapes to the crowd when I glimpsed a man dressed in a long dark coat who stood apart from the rest of the crowd.

It was Billi.

Absorbed in my stories, Billi remained polite throughout the evening and trailed behind at the end of the crowd. Back at the hotel as the tour ended, Billi said, "I like the way you tell stories." Then, like a human-sized crow, he disappeared through the hotel's front doors. I watched through the window as Billi paused at the wait-and-walk sign outside and then crossed the street in the chilled, fall air. Billi attended the ghost tour about two or three times that season. He was always quiet and he was always alone.

In November of 2001, a phone call woke me up about 2 a.m. I recognized the voice. It was the same raspy voice that left a message on my answering machine earlier, with the exact same message: "Hiiiiiiiiii-eeeeeeeee-yaaaaaaaah!" (As in Hi!) The inhuman voice unnerved me so I slammed the receiver down but instantly regretted it. Perhaps I could find out who was pranking me.

Curiosity got the best of me, so I did a *69, which in this area of the country means this will give you the number that last called. I don't remember the exact number but I immediately found out the number was from a Point Pleasant, West Virginia line. I only know two residents of Point Pleasant well enough to call me on the telephone: Jeff Wamsley of the Mothman Museum (and festival) and my brother-in-law's sister. The number belonged to neither of them. I decided I would risk being rude so I dialed the number even though it was past 2 a.m. All I had to do was punch the #1 on my phone as a call back and the phone would ring. I must admit, I was not surprised when an electronic voice clicked on and said, "The number you have called has been disconnected and is no longer in service." Yet, the number had called me five minutes earlier.

I sat down in my chair, shaken. There were perceptions in the back of mind I knew I had been repressing because I considered them impossible. I thought of the appearances of Billi and how he said he had been "born in November of 1966." And how Billi fit perfectly the description of Indrid Cold, with his long dark coat, his tanned complexion and hair combed straight back. And if so, Cold had not aged and still remained at 35 years old. I also remember how diligently Billi had listened to the Woodrow Derenberger tapes, as if he had a secret but never commented. However, this might all be just an overwhelming coincidence so I left it at that.

On Thanksgiving weekend of each year, the local Smoot Theater features a production of *A Christmas Carol* that is put on by the Missoula Theater of Nebraska. It's a grand show which local people look forward to attending year after year. My family (at the time) was no exception. We looked forward to attending on the Friday night presentation in 2001 following Thanksgiving which, as always, was a Thursday. My family had good seats, and we waited as other members of the audience filled the theater. To my right, was my then husband and to my left was an empty chair which was an aisle seat. While my eyes scanned the darkened theater, I then heard a slick rustling and noticed someone sat down beside of me. I turned.

It was Billi. Still wearing his long black coat, I'd never seen Billi change into anything else.

As I sat there, I wanted to fill my husband to what was going on…but I was simply too frightened to. Billi turned to me and said in his naïve way, "I didn't think I was ever going to get here." I nodded, but said nothing. Soon dancers and singers in costume filled the stage. Billi leaned in and whispered, "What do you call what they are wearing up there?"

I finally answered, "Those are their costumes."

Billi commented with a naive amazement, "Aren't those colors beautiful?" I nodded.

The dark-dressed man, who now seemed almost a friend, squirmed in his seat and before the play was over, Billi left. After the play had concluded, my 10 year old daughter commented that she noticed Billi's black coat was made out of an unusual material like the waterproof fabric on a tent.

2001 Holiday Surprise

Holiday shopping and wrapping of presents means lots of pizza deliveries to my home in December. One evening of gift-wrapping brought a surprise to the door. We had ordered a pizza for dinner, decided to decorate the tree and then maybe watch Christmas movies.

When the pizza man knocked on the door, I opened it to find Billi standing there holding a pizza box. I handed him a $20 bill, told him to keep the change, and without acknowledging that I knew him, I quickly slammed the door and locked it. By this time, I had lost my appetite so I didn't eat the pizza and said nothing about it to my family. My family survived the pizza and I survived another surprise visit from Billi.

Around this same time period, my best friend called me and said three of her grandsons had Chiari Syndrome and two of them would need brain surgery. The family would have to travel out of the area for the boys get it. I could not help remember Billi's far-fetched comments about assisting brain surgeries only a few months earlier.

Otherwordly Art Auction

Local art association Artsbridge used to put on what was called their "Chair Auction" where area artists painted used furniture that was auctioned off to help their organization with funding in art education. I was one of the local artists invited to paint a piece of furniture, which I did. I painted an old telephone stand with a Halloween theme and even added a ceramic vintage pumpkin which looked like a dreaming, sleeping child.

This would have been late summer of 2002. On the night of the auction, we assembled at the local Art Center to enjoy the refreshments and watch our art pieces being auctioned off. I was sitting with my friend whose grandsons had just had brain surgery for their Chiari Syndrome. It was not long after 9/11 and everyone needed a lift.

As the second floor banquet room of the Art Center began to fill with people, I noticed Billi in his dark coat enter the room. But he wasn't alone. Billi arrived with a surreal-looking blond companion over six feet tall with pointy, cone-like breasts. Wearing a shade of bright coral lipstick, the woman resembled a late 50s to early 60s pin-up model. She was dressed mostly in black. I poked my friend and commented, "That's Billi!" I pointed in their direction. My friend's eyes widened and you would have thought she was looking at the chain-rattling ghost of Ebenezer Scrooge. But it was Billi all right and he wasn't a ghost.

The auction began. Soon, my art piece came up for bids. The bidding grew fast and heated. My friend's eyes darted back and forth and then her gaze fixed on me. "He's bidding on you," she whispered. (I dared not look.)

Not surprising, Billi was the highest bidder and bought my Halloween-themed, vintage telephone stand. I was stunned for the third or fourth time. When was this going to end? My friend had the composure to go downstairs where items were being paid for, so she followed Billi and his blond bombshell to the check-out area. My friend came back upstairs and said to me, "Susan, Billi paid for your art in cash. He then clasped the stand to his chest and ran out of the Art Center as if he scored some great prize."

After this, I never saw Billi again. It makes me wonder if my art is now being looked upon with quizzical stares by humanoids dressed in black, far, far away in a galaxy called Ganymede, on a planet named Lanulos, perhaps a place that is "less powerful that yours," one inhabited by a guy called "Billi" or maybe even "Indrid Cold." I don't know...However, the strangeness that visited my life wasn't over quite yet...

Continued Phone Calls and Men in Black Appearances

If you have read John Keel's *The Mothman Prophecies* book or watched the movie starring Richard Gere, you already know that both Indrid Cold and the Men in Black have a fascination with phones. Their calls are usually creepy, or seem to be a way of letting us know they are watching us and are aware of what we are saying or doing.

Haunted Parkersburg tour guide Virginia Lyons attended a Mothman Festival with me a year or two after the Billi appearances. Virginia, at the time, was a dedicated home health aide. She had never had any Men in Black experiences, but she was a small child when the Mothman appearances happened, as well as the Woodrow Derenberger UFO encounter and remembered watching his live television interview. Virginia was also the neighbor of Woodrow Derenberger's niece, who lived across the street from her and her husband Steve.

Held in mid-September, the annual Mothman Festival occurs in what is usually the last hot weekend of the year. West Virginia's seasons change quickly but the Mothman Festival is typically sweltering, and making it hard on the actors portraying MIBs who walk around in black suits that are much too hot for the season. I've been friends with paranormal author Rosemary Ellen Guiley for 20 years and we often have dinner together during the festival.

We three were having a late Saturday afternoon dinner at the local Binegar's Restaurant when it was interrupted by phone calls. Virginia had just purchased a new cell phone which she used to keep tabs on her patients. One was having some problems at the time, and even though Virginia had not yet figured out all of the bells and whistles on her new phone, she kept it close by for emergencies.

After we ordered our dinner, Virginia's phone rang. She picked it up. No one spoke but the phone gave off a muffled, underwater sound. She put the phone down and within minutes, it rang again. The call back number was 111-111-1111, one she'd never heard of nor did it exist except with some weird hoaxes involving the *New York Times* in 2011. But this was well before 2011.

Because I'd just given a talk on Indrid Cold and the Men in Black, we guffawed and joked. I commented, "It must be Indrid Cold. He knows we're here." The phone rang a few more times and Virginia tried to check her voice mail but she hadn't figured it out yet. The evening was otherwise uneventful and on Sunday, Virginia and I returned to Parkersburg after another fun-filled Mothman Festival.

The next day Virginia called me. She had figured out her new phone. Someone had left a voice mail for her and she warned me, "Sit down. You are not going to believe this. Let me play it for you." She then played her message and once again there was the strange, rasping voice I recognized saying "Hiiiiiiiiii-eeeeeeeee-yaaaaaaaah" on Virginia's voice mail. It was the same inhuman voice, which again, we took to mean "Hi." But behind that "hi" there seemed to be a message that was meant to unnerve and possibly threaten us, the listeners.

After that, I stopped speaking about Indrid Cold and the Men in Black on the Haunted Parkersburg Ghost Tours, not because

I was scared but because it seemed people would rather listen to ghost stories, and even though the Indrid Cold story happened mostly in the Parkersburg area, the interest did not seem to be there for the strange UFO tale any longer. I also think a part of me didn't want to see Billi again and I feared he might come back.

But the Men in Black weren't done with me yet.

They Arrive, Once Again

In 2006, my mother had an operation on her esophagus that went terribly wrong. She lingered in the hospital and nursing homes until 2009 when she finally died. During her time in the hospital, my sister often stayed with me.

One afternoon while I was napping, my sister heard a rap on the front door. She opened it to see two men dressed in black who seemed eager to talk for some reason. Thinking they were fundamentalist Christians, or holy rollers, from a local church, my sister asked them what they wanted and the men finally asked, "Is Susan at home?"

My sister answered, "Yes, but she is upstairs sleeping." The two men dressed in black didn't seem satisfied with her answer and appeared to crane their necks to look inside the house as if they were searching for some glimpse of me.

They stood for a few moments in awkward silence. One looked at the other and then back at my sister.

"Susan is sleeping?" he asked.

My sister said, "Yes, our mother is sick and Susan is very tired."

The men again seemed to grasp for words. As my sister looked behind them, she saw two other men dressed in black across the street, standing as if they were waiting on the two men staring in the house. Finally, the one man at the door said, "Please tell Susan we were here. We hope she feels better. We will see her again."

The two men dressed in black walked down the front cement steps to my house. They joined the remaining two men, got into a non-descript black car and all four men who were dressed in black drove away without any fanfare. They did not seem to be interested in visiting any other home.

That was ten years ago and was my last encounter with any Men in Black. To my relief, Billi has never reappeared. I would think if he had been just a local resident, I would have seen him again over time. But I haven't. Billi vanished from my life as mysteriously as he arrived. However, since I plan to add the Men in Black and Indrid Cold to the fall Haunted Parkersburg tours in 2017 by playing the old UFO tapes, it is very likely they will somehow make their presence known in my life again. Frankly, I'd be surprised if they didn't.

26

"SHE DID NOT MOVE OR SPEAK. SHE JUST STARED"

October 14, 201

Dream invasions and missing government documents

If you want weird, you've got it. It didn't get much more bizarre than Saturday morning, October 14. The night before, I gave a lecture in Fort Worth, Texas for a local group called Satori. It was a good turn out and among those who turned up was good friend, Kimberly Rackley, who, has had her very own run-ins with the MIB. During the lecture, I made mention of how certain supernatural things can invade our dreams – and, in the process, turn those same dreams into absolute, mind-warping nightmares. I guess someone – or *something* – was tuning in, as the following morning I received at Facebook Message a pair of strange tales of – wait for it – Men in Black and Women in Black who had invaded the dreams of the victims. No, I do not jest.

The first came from a guy in Florida. I had spoken with him before. In my 2016 book, *Women in Black*, I told the very strange story of "A Hesitant Believer," who had a very chilling encounter with both a MIB and a WIB a few years earlier. It was a hostility-filled encounter, in which the clearly non-human pair in black put a hex on poor "AHB," one that left him extremely ill for a while. On this date, though, he wanted to tell me of a follow-up encounter, which was no less strange or unsettling. He told me:

> Mr. Redfern, I wrote to you a couple of years ago about a possible WIB/MIB encounter in Tampa, FL. I noticed you recently posted my account on *Mysterious Universe*, and that reminded me to tell you of an incredibly vivid nightmare/sleep-paralysis episode I had in late December 2016 which involved the WIB. This "dream" has been the only one I have ever had about my experience. I had meant to write to you earlier, but life got in the way, and I didn't want to bother you with what may have been a perfectly ordinary night terror. If such 'follow-ups' are relevant to your research, let me know. Sincerely.

A bit stunned that "AHB's" message should have arrived barely twelve hours after I had been speaking about dream-invaders before the Satori audience in Fort Worth, I replied quickly: "Many thanks for this! Yes, I'd be very pleased to hear about the dream, as I have quite a few cases on file where people had seen the MIB in their dreams, and as if the MIB can 'invade' our dreams and even manipulate and control our dreams too. Best, Nick."

"AHB" expanded:

> Ok. Thank you again for taking this seriously. This dream was very traumatic, but I will try to describe it the best I can. A bit of background: I had moved in with my disabled parents in order to take care of them. The date of the "dream" was December 21st, 2016. I remember this because my best friend's birthday was the next day, and we'd planned to make a day of it, so I went to bed early. I had not been drinking or doing drugs, etc. My parents

were out of the house and staying with my aunt, and I was alone.

I drifted off, and when I "awoke", I found myself in a state of sleep paralysis. I've had these episodes since I was a child, but I'd never seen or felt any other presences during them. This time my bed was surrounded by pinkish-orange flames that took the shape of what I would describe as "goblins." At the foot of my bed loomed a large hand, larger than my bed, and wearing a tattered suede glove.

I was terrified at first, but when I realized I couldn't move, I knew this was just a dream and I was suffering from sleep paralysis, although I was seeing "entities" for the first time. To try to explain it, I will say my mind was logical but my body was frightened, if that makes any sense at all. I tried to just ride it out and hope I would fall back asleep soon, but then I noticed a shifting from by the bedroom door.

The woman I had encountered in the bar [the story of which is told in *Women in Black*] was standing there, but she was dressed differently. She did not have the bright green scarf and was wearing a conservative black dress, like something a religious woman would wear to a funeral. No hat or veil, though. She had that same short hair with bangs that looked like a bad wig, and around her neck was a chain with a pendant that resembled a coin or medallion, which I do not recall her wearing in the bar [Author's note: oddly, coins and medallions turn up in a lot of MIB-themed cases].

When I saw her, my mind began to panic, and I am sorry, but I just cannot go into detail about how I felt at that point. I still realized that the goblins and flames and the hand were part of the dream, but I knew the woman was

> real. She was actually standing there, at least according to my mind. She did not move or speak. She just stared. Then the dream elements disappeared, but she was still there.
>
> For some reason, this was the most horrifying part: she didn't disappear, or flash out, or do anything supernatural, but calmly turned, opened my door, and left. I heard the front door open and the sound of a car engine. I lapsed back into sleep after that. When I woke up, it was morning, and I was incredibly nauseous. I vomited, and needless to say cancelled plans with my friend. Made me feel like an ass, but I couldn't explain it to him. Anyway, that was my dream. It actually took a lot out of me to write that, and I hope it's sufficient, but if you have any questions, just ask. I'm hung up about the coin/medallion pendant the woman was wearing. I just felt this overwhelming urge not to look at it.

Then, as I scrolled down my messages, there was this one from a girl named Emily:

> Hello Nick! My name is Emily. I recently read an article you posted about a man in a black cloak, as I was researching a very scary dream I had of this figure last night. Something that stood out in the article was that the other women that encountered this were in their early 20s and have had alien type abduction encounters, another thing I have experienced in the past. Was hoping you might have any evolved info on this since that original posting. I am extremely interested in this phenomenon and would like to know more thanks!

I replied:

Hey Emily, Thanks for this! I actually have a lot of reports of Men in Black and Women in Black where the encounter occurs in a dream. I don't think these are normal dreams though. There is good evidence that the MIB and the WIB can 'invade' peoples' dreams, and become part of the dream (or nightmare) and control and manipulate the dream. Yes, a lot of people in their teens and twenties who have had these experiences have had UFO experiences, or have been involved in paranormal phenomena, such as Ouija boards. I have many reports of 'dream invasions' by the MIB / WIB and they are all around the world. The numbers appear to be growing. If you want to chat about this, I can give you my phone number. Best, Nick.

Emily chose not to call, but she added:

> That is very interesting because I was definitely shaken when I read that about the UFO experiences. I had a really realistic alien dream a while ago and then this scary dream about the man in black and he was wearing a grim reaper cloak that was billowing with wind. And he set my car on fire in the dream.

And, if that were not enough strangeness for one day, as I was getting close to finishing the writing for the book you are now reading, I decided to make a start on the photo section of the book – this was on the same day, October 14. I figured that it would be cool to reproduce one of the documents from the FBI's *Challenger* Space Shuttle file – the file which provoked so much intrigue when my agent, Lisa Hagan, got a weird phone call from someone who simply said two words: "*Challenger* exploded."

So, I went to the FBI's website, The Vault. The file was still there, split into three PDF files. I downloaded them, as I was unsure which of the three files the specific documents were in. I searched the first one. No luck. Ditto for the second. And for the third too. Then, I scrolled through the whole document collection again. And once again. The combined file was intact – aside from the pages that dealt with the strange saga of the Space Shuttle, which I tell in Chapter Three of this book. Someone in the FBI had clearly removed the relevant pages from the PDF. For what reasons? What was the motivation? Who gave the order? And why? To this day, I still don't know the answers to those questions. What I *do* know, though, is that this was sure as hell one of the weirdest ways to wake up on a Saturday morning! I told Lisa a couple of days later (there was no point in upsetting her weekend). She freaked.

And which is, I think, a good place to bring this strange, three-years-plus saga to its end.

CONCLUSIONS

At the time I write these final words, it's January 12, 2018.. Looking back on the events described in the pages of this book, I cannot deny that it has been a strange, uncanny and – at times – even dangerous, journey. But, that's how things go when you are up against the Men in Black. Across three books - *Men in Black, Women in Black* and now *The Black Diary* - I have detailed the turbulent and terror-filled encounters of numerous witnesses to the MIB and WIB phenomena. And, I have revealed to a stark degree the way in which that same phenomenon has impacted on me, too. Also, let's not forget my other, earlier books on the MIB: *The Real Men in Black* and *On the Trail of the Saucer Spies.*

So, in light of all the above, what can I say in a few, concluding words? Well, there is absolutely no doubt in my mind – at all – that the Men in Black phenomenon is all too real. But, it has nothing to do with government personnel, "secret agents," or the likes of Will Smith's and Tommy Lee Jones' characters "J" and "K," in the three phenomenally successful *Men in Black* movies. The true MIB are hostile, malevolent, manipulative things that can invade our dreams, provoke bad luck and negativity – and even cause illness and plunge our minds into states of near-madness and unbridled paranoia. They are to be avoided at all costs and whenever and wherever possible.

Whatever the real MIB (and their various, motley offshoots) are, all I can say with certainty is that they are not human. Demons? Time-Travelers? Tricksters? Multi-dimensional creatures? The denizens of some terrible domain that we cannot fully understand? The theories are many. The firm answers, unfortunately, elude us, despite the wealth of testimony in support of their reality.

I can't say for sure that my time spent investigating the MIB controversy is completely over. It probably isn't. Indeed, I would not be surprised if, one day, my path again crosses theirs. For now, at least, my journey is at its end. It has, to be sure, been a wild and crazy ride.

If my words – spelled out across a series of books filled with supernatural high-strangeness – have convinced you to steer clear of the MIB and the WIB, then that's not a bad thing. I suspect, though, that in the same way that I have found it so difficult to walk away from the phenomenon, you too might find yourself caught up in a maelstrom from which there is little or no escape. For your sake, I hope it's the former. I suspect though, that it just might be the latter. Whichever path you decide to negotiate, take care.

BIBLIOGRAPHY AND SUGGESTED READING

"Albert K. Bender." https://en.wikipedia.org/wiki/Albert_K._Bender. 2017.

Barker, Gray. *Bender Mystery Confirmed.* Clarksburg, WV: Saucerian Books, 1962.

Barker, Gray. *M.I.B.: The Secret Terror Among Us.* Jane Lew, WV: New Age Press, 1984.

Barker, Gray. *They Knew Too Much About Flying Saucers.* Clarksburg, WV: Saucerian Press, Inc., 1975.

Barker, Gray. *When Men in Black Attack.* Point Pleasant, WV: New Saucerian Books, 2015.

Beckley, Timothy Green. *The UFO Silencers.* New Brunswick, NJ: Inner Light Publications, 1990.

Bender, Albert. *Flying Saucers and the Three Men.* NY: Paperback Library, Inc., 1968.

Berlitz, Charles & Moore, William. *The Philadelphia Experiment.* St. Albans, U.K.: Granada Publishing Limited, 1979.

Constable, Trevor James. *They Live in the Sky.* Watertown, MA: New Age Publishing, Co., 1958.

"Crypto-Kid." https://www.facebook.com/paranorm101/. 2017.

Dewey, Steve & Ries, John. *In Alien Heat.* San Antonio, TX: Anomalist Books, 2005.

Godfrey, Linda. *Monsters Among Us.* NY: Tarcher Pedigree, 2016.

Graaf, Mia De & O'Hare, Sean. "EXCLUSIVE: Screenwriter mysteriously killed in 1997 after finishing script that revealed the 'real reason' for US invasion of Panama had been working for the CIA... and both his hands were missing."

http://www.dailymail.co.uk/news/article-2905392/Hollywood-screenwriter-mysteriously-killed-20-years-ago-working-CIA-hands-sent-autopsy-200-years-old.html. January 17, 2015.

Guiley, Rosemary Ellen. *The Djinn Connection.* Pensacola, FL: Visionary Living, Inc., 2013.

Guiley, Rosemary Ellen & Imbrogno, Philip J. *The Vengeful Djinn.* Woodbury, MN: Llewellyn Publications, 2011.

Hatfield, Samuel. "Dream Invasion." https://samuelhatfield.com/articles/dream-invasion.html. December 6, 2012.

"Haunted Parkersburg Ghost Tours." https://www.hauntedparkersburgtours.com/. 2017.

Hill, Brian. "Lovelock Cave: A Tale of Giants or a Giant Tale of Fiction." http://www.ancient-origins.net/myths-legends-americas/lovelock-cave-tale-giants-or-giant-tale-fiction-003060. May 15, 2015.

Hollis, Heidi. *The Hat Man.* Milwaukee, WI: Level Head Publishing, 2014.

Jessup, Morris J. *The Case for the UFO.* NJ: Citadel, 1955.

Keel, John. *The Mothman Prophecies.* NY: Tor Books, 2002.

Keith, Jim. *Casebook on the Men in Black.* Lilburn, GA: IllumiNet Press, 1997.

Levin, Ira. *Rosemary's Baby.* NY: Pegasus Books, 2017.

Medway, Gareth J. "Men in Black Encounters, a Short Catalogue." http://pelicanist.blogspot.com/p/mib-encounters.html. 2017.

Microwave Techniques. MA: Department of the Navy, May 1, 1950.

"Neil Armstrong." https://vault.fbi.gov/neil-armstrong. 2017.

Nelson, David. "Mysterious 'men in black' sightings reported along Muscatine Co. roadways." http://www.kwqc.com/content/misc/Mysterious-men-in-black-sightings-reported-along-

Muscatine-Co-roadways-415398733.html. June 21, 2016.

Offutt, Jason. *Darkness Walks.* San Antonio, TX: Anomalist Books, 2009.

"Philadelphia Experiment." https://www.history.navy.mil/research/library/online-reading-room/title-list-alphabetically/p/philadelphia-experiment/philadelphia-experiment-onr-info-sheet.html. September 2, 2016.

Randles, Jenny. *The Truth Behind the Men in Black.* NY: St. Martin's Paperbacks, 1997.

Redfern, Nick. "Black Helicopters: What in Hell?!"

http://mysteriousuniverse.org/2016/11/black-helicopters-what-in-hell/. November 2, 2016.

Redfern, Nick. *Men in Black.* VA/NY: Lisa Hagan Books, 2015.

Redfern, Nick. *On the Trail of the Saucer Spies.* San Antonio, TX: Anomalist Books, 2006.

Redfern, Nick. *The Real Men in Black.* Wayne, NJ: New Page Books, 2011.

Redfern, Nick. *The Slenderman Mysteries.* Wayne, NJ: New Page Books, 2018.

Redfern, Nick. *Women in Black.* VA/ NY Lisa Hagan Books, 2016.

Reynolds, Rich. "Are Nick Redfern's WIB witnesses nuts? https://ufocon.blogspot.com/2016_06_22_archive.html. June 22, 2016.

Reynolds, Rich. "Nick Redfern's latest Book: The Women in Black." https://ufocon.blogspot.com/2016_06_20_archive.html. June 20, 2016.

"Space Shuttle Challenger." https://vault.fbi.gov/Space%20Shuttle%20Challenger%20. 2017.

"Star and Crescent of Islam." http://symboldictionary.net/?p=2395. 2017.

Stuart, John & Beckley, Timothy Green. *Curse of the Men in Black.* New Brunswick, NJ: Global Communications, 2010.

"The Tracie Austin Show. http://thetracieaustinshow.com/. 2017.

Weatherly, David. *Strange Intruders.* Denton, TX: Leprechaun Press, 2013.

Weatherly, David. *The Black Eyed Children.* Denton, TX: Leprechaun Press, 2012.

ACKNOWLEDGMENTS

I would like to offer my sincere thanks to my agent, friend and publisher, Lisa Hagan; my editor and co-publisher, Beth Wareham; and book-designer, Simon Hartshorne. And, of course, a *very* big thank-you to everyone who contributed to this book and who helped to make its publication possible.

ABOUT THE AUTHOR

Nick Redfern is the author of more than 40 books on UFOs, lake-monsters, zombies, and Hollywood scandal, including *The Roswell UFO Conspiracy*; *Women in Black*; *Men in Black*; and *365 Days of UFOs*. Nick has appeared on many TV shows, including the BBC's *Out of This World*; the SyFy Channel's *Proof Positive*; the History Channel's *Monster Quest, America's Book of Secrets, Ancient Aliens* and *UFO Hunters*; the National Geographic Channel's *Paranatural*; and MSNBC's *Countdown* with Keith Olbermann. Nick lives in Arlington, Texas.

He can be contacted at his blog: http://nickredfernfortean.blogspot.com

Made in the USA
Coppell, TX
26 June 2021

58090574R10144